New Analytical Methods in Earth and Environmental Science

A new e-book series from Wiley-Blackwell

Because of the plethora of analytical techniques now available, and the acceleration of technological advance, many earth scientists find it difficult to know where to turn for reliable information on the latest tools at their disposal, and may lack the expertise to assess the relative strengths or potential limitations of a particular technique. This new series addresses these difficulties, and by virtue of its comprehensive and up-to-date coverage, provides a trusted resource for researchers, advanced students, and applied earth scientists wishing to familiarise themselves with emerging techniques in their field.

Authors will be encouraged to reach out beyond their immediate speciality to the wider earth science community, and to regularly update their contributions in the light of new developments.

Written by leading international figures, the volumes in the series will typically be 75–200 pages (30,000–60,000 words) in length—longer than a typical review article, but shorter than a normal book. Volumes in the series will deal with

- the elucidation and evaluation of new analytical, numerical modeling, imaging, or measurement tools/techniques that are expected to have, or are already having, a major impact on the subject;
- new applications of established techniques;
- interdisciplinary applications using novel combinations of techniques.

All titles in this series are available in a variety of full-colour, searchable e-book formats. Titles are also available in an enhanced e-book edition, which may include additional features such as DOI linking, high resolution graphics, and video.

Series Editors

Kurt Konhauser, University of Alberta (biogeosciences)
Simon Turner, Macquarie University (magmatic geochemistry)
Arjun Heimsath, Arizona State University (earth-surface processes)
Peter Ryan, Middlebury College (environmental/low T geochemistry)
Mark Everett, Texas A&M (applied geophysics)

GROUND-PENETRATING RADAR FOR GEOARCHAEOLOGY

LAWRENCE B. CONYERS

WILEY Blackwell

Library of Congress Cataloging-in-Publication Data

Conyers, Lawrence B., author.
Ground-penetrating radar for geoarchaeology / Lawrence B. Conyers.
    pages   cm
  Includes bibliographical references and index.
  ISBN 978-1-118-94994-8 (cloth)
1. Ground penetrating radar.   2. Archaeological geology.   I. Title.
  TK6592.G7C66 2016
  930.1028–dc23
                                        2015020391

A catalogue record for this book is available from the British Library.

Cover image: Amplitude slice map across a sand dune sequence from coastal Brazil showing occupational surfaces on the bottom, cross-beds within the dunes in the middle, and highly reflective inter-dune sediments at the top © Lawrence Conyers

Set in 9.5/11.5pt Minion by SPi Global, Pondicherry, India
Printed and bound in Singapore by Markono Print Media Pte Ltd

1   2016

Contents

About the Author

Lawrence B. Conyers is a professor of anthropology at the University of Denver, Colorado. He received a Bachelor of Science degree in geology from Oregon State University and a Master of Science degree from Arizona State University. He holds both M.A. and Ph.D. degrees in anthropology from the University of Colorado, Boulder. Before turning his attention to ground-penetrating radar and other near-surface geophysics for archaeological mapping, he spent 17 years in petroleum exploration and development where he worked with seismic geophysical prospecting. His GPR research is conducted throughout the United States and at many sites throughout the world.

Acknowledgments

I am fortunate very early in my professional life to have worked with excellent geophysicists who helped me along with seismic reflection interpretation during the years I was in the petroleum business. Randy Ray, Phil Howell, and Bill Miller were all instrumental in my early growth in the subject. Bill Miller actually showed me the very first GPR profile sometime in the 1980s when GPR was very new to both of us. I remember talking to him about how we could immediately appreciate ways this new device could be interpreted in much the same way as seismic reflection. As it turned out, many of the original GPR processing programs were taken directly from seismic work, so it was natural to fall right into GPR. I was gladdened to see in the research for this book that my professor of geophysics in graduate school back in the 1970s, Bill Sauck, has now started to do GPR research, and I have cited a recent publication by him and one of his graduate students in Chapter 6. When I was in graduate school I was also fortunate to have Payson Sheets as an advisor, who thinks like a geologist and geophysicist as well as an archaeologist. While studying with Payson I was very lucky to also be advised on soils and geomorphology by Peter Birkeland, who was instrumental in getting me to think about near-surface sediments and soils in ways that were directly applicable to GPR and archaeology. None of my work with GPR could have been accomplished without the intelligent collaboration and loyal friendship of two true geniuses of GPR research, Jeff Lucius and Dean Goodman.

Early on in my academic career I was encouraged to pursue GPR and geoarchaeology research at the University of Denver, even though many of my colleagues in the Anthropology Department considered these subjects to be perhaps a little "too scientific" or at least very esoteric. They were always encouraging and supportive, and with other geological friends across campus in the Geography Department, Don Sullivan and Mike Daniels, I had my own "team" of collaborators to work with on a number of great projects, some of which are used as examples in this book.

Many other collaborators in projects around the world have been always easy to work with and tolerated much trial and error on my part within the framework of their own research. All were great colleagues, and most of the examples provided here would not have come about without their generous and supportive collaboration.

My many students, both graduate and undergraduate, did much of the hard work in data collection. They asked all the right questions to keep me on track, and provided many of the obvious answers when I was flummoxed by complex datasets. A list of these many wonderful coworkers, collaborators, students, and colleagues who helped provide the data, insights, and hard work necessary to provide the examples for this book includes, but is not limited to, Melissa Agnew, Peter Arceo, John Arthur, Tiago Attore, Andrew Bauman, Robert Bauman, Leigh-Ann Bedal, Mike Benedetti, Lucia Bermejo, Nuno Bicho, Federica Boschi, Tamara Bray, Charles Bristow, Richard Buckley, Carmen Buttler, Scott Byram, Maria Caffrey, James Conyers, Anika Cook, Lauren Couey,

Matthew Curtis, Mike Daniels, Mike Disilets, Kristin Elliott, Eileen Ernenwein, Paul Fish, Suzanne Fish, Nicole Fulton, Matt Golsch, Dean Goodman, Jonathan Haws, John Hildebrand, Maren Hopkins, Sally Horn, Gary Huckleberry, Harry Jol, Christina Klein, Michele Koons, Jeff Lucius, Harold Luebke, Cody Main, Corey Malcolm, Ana Isabel Ortega Martinez, Ian Moffat, Jeremy Moss, Dani Nadel, Steve Nash, Samantha Nemecek, Matt Pailes, Lisa Piscopo, Hannah Polston, Yossi Salmon, Payson Sheets, Zach Starke, Jennie Sturm, Don Sullivan, Mary-Jean Sutton, Tiffany Tchakirides, Victor Thompson, Chet Walker, Lynley Wallis, Katrina Waechter, and Sean Wiggins.

Introduction to Ground-penetrating Radar in Geoarchaeology Studies

Abstract: Geology and archaeology have long been integrated as a way to understand site formation processes, place artifacts within an environmental context, and as a way to study ancient people within the landscapes where they worked and lived. An analysis of sedimentary environments has long been necessary in this endeavor, but is often constrained by a lack of excavations, exposures, and other data to study areas in a three-dimensional way. Ground-penetrating radar (GPR) has unique three-dimensional abilities to place ancient people into an environmental context by integrating both archaeological and geological information within the buried context of a site over wider areas that is usually possible. The GPR method can accomplish this because it is based on the analysis of reflections produced from the interfaces and layers of geological units in the ground that are then studied three dimensionally. When this is done, robust analyses of buried geological and archaeological materials can be done for subsurface areas not visible at the surface in order to generate more holistic analyses of geoarchaeological studies.

Keywords: environmental context, sedimentary environments, three-dimensional analysis, buried materials and strata, stratigraphy, reflection generation, environmental reconstruction

Introduction

There has been a long period of collaboration between geologists and archaeologists, as it is impossible to separate the geological record from archaeological materials preserved within sediments and soils. These cross-disciplinary geoarchaeology studies involved stratigraphic analysis, environmental reconstructions, site selection for future excavations, and an analysis of site preservation and postabandonment processes (Butzer 1971; Rapp and Hill 2006). More recently, these types of collaborative geological and archaeological studies have included landscape analysis that places people within an often complex and changing environment (Bruno and Thomas 2008; Constante et al. 2010; Stern 2008). The inclusion of geophysical analysis within geological and archaeological studies has occurred more recently and is beginning to make an impact in many research projects (Campana and Piro 2008; Kvamme 2003) as buried deposits can be studied and integrated with more limited excavations and exposures. These geophysical studies for the most part employ magnetics, electromagnetic induction and electrical resistivity, and ground-penetrating radar (GPR). The use of these types of geophysical methods allows

Ground-penetrating Radar for Geoarchaeology, First Edition. Lawrence B. Conyers.
© 2016 John Wiley & Sons, Ltd. Published 2016 by John Wiley & Sons, Ltd.

a more complete and broader aerial analysis of complex buried (and otherwise invisible) archaeological and geological materials than was possible in the past (Johnson 2006).

This book is devoted to one of these geophysical methods, GPR, and especially the integration of its unique imaging properties to measure and display materials in the ground along with geological and archaeological data. The GPR method transmits radar (electromagnetic waves) energy into the ground and then measures the elapsed time and strength of reflected waves as they are received back at the ground surface (Figure 1.1). Many thousands or hundreds of thousands of reflected waves are collected along the transects of antennas as they are moved along the ground surface to produce reflection profiles of buried layers and features analogous to viewing profiles in excavation trenches (Figure 1.2). When many reflection profiles are collected in a grid, three-dimensional images of buried materials in the ground can be constructed (Conyers 2013, p. 166). Ground-penetrating radar therefore has the unique ability to not just produce images of both geological and archaeological units in the ground, but to do so in three dimensions (Conyers 2012, p. 20).

Ground-penetrating radar's ability to produce two- and three-dimensional images of soils and sediments within depths that are usually of importance for archaeology (a few centimeters to 3–4 m burial at most) means that complex images of geological materials associated with archaeological deposits is possible. While some archaeological thinking views the geological matrix of a site as a volume of material that must be removed and discarded to get to the important artifacts and features, most recognize that there is important information to be gained by studying it (Davidson and Shackley 1976; Waters 1992, p. 15). It is this appreciation that geology cannot be divorced from archaeological research that forms the basis for the field of geoarchaeology. This cross-disciplinary focus can become even more important when GPR is integrated with the other datasets to project important information from the visible areas in outcrops or excavations into the invisible and still buried areas of a site.

Often much of what can be seen in GPR profiles and three-dimensional amplitude maps is more geological than archaeological, and there can often be confusion as to what is anthropogenic in origin, or instead the geological matrix (Conyers 2012, p. 19). Successful

Control system and display monitor

Distanced encoder wheel   270 MHz antennas

Figure 1.1 Collecting GPR profiles with a GSSI SIR-3000 control system and 270 MHz antennas.

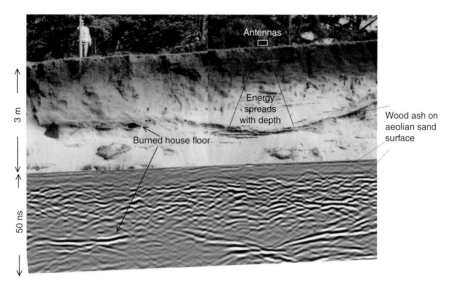

Figure 1.2 Comparison of a 400 MHz reflection profile collected within a 50 ns time window to a 3 m thick outcrop of cross-bedded aeolian dune sands with a burned house floor. Reflection energy spreads from the surface transmission antenna, creating an average of reflections received back at the ground surface from subsurface interfaces. From Conyers (2012). © Left Coast Press, Inc.

differentiation of the two, and an interpretation of radar **reflections** derived from all the units in the ground, is therefore crucial. As most archaeological sites are the result of burial and preservation by geological forces and processes, the various features in the ground that have been modified and altered by physical and chemical forces must be understood. This can be difficult even when exposures are visible to the human eye, but especially challenging when various buried features are visible but not necessarily understood in GPR images. The application of GPR to both geological and archaeological features and their interpretation within standard GPR-processed images is the goal of this book.

Scales and Applications of Geoarchaeological Studies with GPR

Geoarchaeological studies range in scale from very small scale analysis of micromorphology of soils and sediments using the microscope to large landscapes covering huge tracts of land (Goldberg et al. 2001; Rapp and Hill 2006). The GPR method of acquisition and data processing methods has very specific resolutions at measurable depths, which necessitates that it be employed within a middle-range of the usual standard geoarchaeology studies. These scales of study typically involve a few hectares aerial extent at most, with depth of analysis of 3–4 m and feature resolution usually larger than about 20 cm in the maximum extent. There are some notable examples of very large data sets recently collected by **multiple array systems** towed by motorized vehicles that can study many tens or even hundreds of hectares (Gaffney et al. 2012; Trinks et al. 2010) but these are still relatively rare. Within the scope of most geoarchaeological applications (French 2003, p. 6), and with most of the examples presented here, the study area may be on the order of a few hundreds of square meters in dimension to depths of about 6–7 m.

Geological analysis within the context of archaeology, which can be expanded on and broadened using GPR, can be used to study landscape evolution (Ricklis and Blum 1997) where settlement changes are a function of environmental fluctuations. Specifically, GPR datasets can define fluvial units that are the product of erosion and redeposition (Behrensmeyer and Hill 1998) and associated soil units, which are a function of landscape stability over many centuries (Birkeland 1999; Ferring 2001; Holliday 2001). An analysis of these geological units using an integration of stratigraphic units (Shackley 1975) with GPR datasets (Conyers 2009), within a dynamic landscape will also allow for the study of site formation processes (Schiffer 1972).

Studies that are expanded beyond site formation processes can show the effects of humans on a landscape and their adaptation to environmental change over time (Campana and Piro 2008). This is done by focusing on the geological matrix of a site first, defining depositional environments and changes in those environments laterally and vertically over time. The archaeological record is then placed within this context to understand human adaptation to and modification of their environment. This definition and understanding of environments is one of the key foci in GPR integration with geoarchaeology. This book will provide examples of various common environments discernable in GPR data sets, and then place human activities within those contexts.

The important geological packages of sediments and associated geological units that can be studied and analyzed with GPR are most of the terrestrial depositional environments (such as rivers, floodplains, sand dunes, beaches and other coastal environments), bedrock features that were part of an erosional landscape and later buried, and soil horizons that were living surfaces providing some degree of stability in the past. These types of buried features must usually be defined first in excavations and outcrops, and then projected into areas where they are buried and invisible except by using GPR techniques.

A key to understanding past environments is to first define the general stratigraphy of buried units and understand how those units are visible in common GPR images. This is not always as straightforward as would be hoped, as the varying chemical and physical properties of buried materials sometimes allows reflection of radar waves, and at other times does not. Depth of energy penetration, radar wave attenuation, the spreading of transmitted radar waves as they travel in the ground, and a variety of other variables relating to radar wave properties can often confuse and mislead some interpreters. Often these problems are solvable, and many examples regarding resolution, depth of analysis, and interpretation of the results of data processing are included. For the most part the larger scale geological units, and sometimes their associated sedimentary structures, are readily visible with GPR, and these can readily define specific ancient environments. When GPR interpretations are enhanced with subsurface information derived from augering, cores, and small scale excavations, a three-dimensional analysis of broad landscape features and past environments is usually possible. Facies analysis of larger scale geological units can then be integrated with anthropogenic features and sometimes associated soils to place humans within ancient and historical landscapes.

Basics of the GPR Method

Ground-penetrating radar data are acquired by reflecting pulses of radar energy produced from a surface antenna, which generates waves of various wavelengths that propagate downward (**Figure 1.1**). They spread as they move into the ground in a cone (**Figure 1.2**), which is a function of the physical and chemical properties of the materials through which they pass (Conyers 2013, p. 47). As these waves move through the ground they are reflected from buried objects, archaeological features and stratigraphic bedding surfaces. The reflected waves then travel back to the ground surface to be detected and recorded at a receiving antenna, which is paired with the transmitting antenna. The two-way travel times of the waves moving through the ground are measured at the receiving antenna and their arrivals recorded in elapsed time of travel, in nanoseconds. As the propagating radar waves pass through various materials in the ground their velocity will also change, depending on the physical and chemical properties of the material through which they are traveling (Conyers 2013, p. 107). If the constituent differences at interfaces of materials occur abruptly along boundaries between very different materials in the ground the radar waves' propagating velocity will also change when they pass across the contacts. When this occurs a reflected wave is generated that can move back to the ground surface from the reflection interface. Not all radar waves will travel back to the ground surface at a reflection interface and some energy will continue to propagate deeper in the ground to be reflected again from more deeply buried interfaces, until all the energy finally dissipates with depth. Only the reflected energy that travels back to the surface antenna is recorded and visible for interpretation. If buried surfaces that reflect energy are oriented in a way that reflected waves move away from the surface antenna, that energy will not be recorded, making those interfaces effectively invisible using the GPR method.

Reflections generated from radar waves propagating in the ground are created at interfaces where differing materials are in contact along a boundary and are different enough so that the velocity of moving waves that intersect the interface changes abruptly (Conyers 2013, p. 27). An example of a composition change that affects velocity in this way might be where a clay floor rests on an underlying sand bed (Figure 1.2), and where these materials are then buried by a different material. The contacts of the base of the clay floor with the underlying sand as well as the top of the floor covered with different sand are two interfaces that could generate wave reflections. The radar waves propagated from the ground surface antenna would be moving at a fairly rapid rate in the overlying material, slow abruptly as they passed into the clay floor, and move at an increased rate again as they passed out of the clay floor into the underlying sand. Each abrupt velocity change would theoretically create a reflected wave (Conyers 2013, p. 28). In contrast, a gradational change in materials over some distance would not produce a reflection as there would not be any abrupt change in radar velocity and no reflected wave would be generated. This kind of gradational change might be found when the sediment in one layer changes from silt to sand over a distance of a meter or so. In general, the greater the change in velocity across a boundary, the greater the amplitude of the wave that is reflected back to the surface and recorded.

Reflection profiles are the basic interpretive tool for GPR and are created as radar antennas move along the ground surface transmitting waves downward into the ground. A sequential stacking of many hundreds of reflections (termed traces) consisting of reflected waves from different depths in the ground is then produced. Each trace is recorded at a discrete position along an antenna transect, and the display of all these is used to produce a two-dimensional vertical slice in the ground (Figure 1.2). Profiles of reflections are the standard images used for geoarchaeological interpretations of buried materials in the ground. These will be used throughout this book, as they are the tool to identify and understand geological layers as well as archaeological components within those geological packages. Many reflection profiles collected in a grid can also be processed together in order to produce individual maps of various depth slices in the ground and renderings of features in three dimensions.

Integrating GPR with the Geological and Archaeological Record

Usually prior to conducting a GPR survey, there is a basic knowledge of the geological units and human occupation of an area. This kind of background information can be obtained from previous investigations, the published literature, or from others who have worked in the area previously. Without at least a basic understanding of what geological and archaeological materials to expect in the ground, results of a GPR survey would remain speculative at best (Conyers 2012). Only after obtaining this information can a knowledgeable and at least partially informed study commence. It is best to begin by collecting GPR profiles close to excavations or outcrops where exposures of the units of interest can be studied (Conyers 2013, p. 149). In this way, GPR reflection profiles can be "tied" directly to what is visible and known in exposures (**Figure 1.2**). This way of initiating a project is usually quite direct and can yield immediate results, with specific radar reflections generated from buried layers of interest easily defined and understandable. Radar reflection recording times can also be directly compared to depth of units in the ground and the velocity of radar travel times calculated (Conyers 2013, p. 153).

However, this optimal strategy using direct comparisons between the visible and the radar images prior to conducting a broader GPR investigation is often not possible. This could occur when there are no outcrops or excavations available or when time constraints or lack of permission for subsurface testing has not been obtained. The absence of specific geological or archaeological knowledge prior to GPR research need not impede at least initial investigations, if at least a general knowledge of what is expected in the ground is present.

This situation was confronted in coastal Portugal, where excavations to the west of a dry lake uncovered Late Paleolithic artifacts associated with a temporary hunting camp (Conyers et al. 2013; Haws et al. 2011). These artifacts were found in aeolian sand close to an unconformity with underlying Jurassic bedrock and overlain by a Late Pleistocene soil unit (**Figure 1.3**). They were therefore confined within a stratigraphic package that could be defined using GPR profile analysis where these units were visible.

A GPR survey was conducted around those excavations and an ancient stream channel incised into the Jurassic bedrock was found adjacent to the artifact concentration, which flowed from coastal hills just to the west, eastward toward a main river system that drained this area of western Portugal (Conyers et al. 2013). The same age artifacts (Magdalenian) were also found as surface scatters in plowed ground along the margin of the dry lake bed just to the east in what was presumably the ancient floodplain of the larger river. This suggested that these people were exploiting resources along the margin of this ancient lake environment within the floodplain.

With only this general information to help with interpretation a set of GPR profiles were collected within and to the east of the dry lake bed using 270 MHz antennas (**Figure 1.3**). Some exceptionally high-resolution reflection profiles were obtained in this plowed area used for a pine tree plantation, which had recently been harvested and was lying fallow. After these profiles were collected the recorded radar reflections were

Figure 1.3 Location of the GPR reflection profile used to place Paleolithic artifacts into a geological context in western Portugal.

Late Paleolithic artifacts found in plow scars, mixed within aeolian sand

adjusted for surface topography and then displayed after minor background **noise** removal and resetting of the reflection amplitudes so they could be visible (Conyers 2013, p. 134).

In this case all that was known prior to the GPR survey was that an ancient stream system was located to the west of the survey area toward the floodplain. A dry lake bed was visible on the surface, which might have been present in antiquity and was located at the level of the main river floodplain (now a river terrace surface above the modern floodplain). It appears that humans had discarded Late Paleolithic stone artifacts in the general vicinity, but their relationship to the lake or anything present under the ground surface was not known. The artifacts were found out of place within the aeolian sand, which mantles this area today. They were likely moved from their original stratigraphic positions both vertically and horizontally as this area has been greatly disturbed by plowing, planting of trees for pine pitch extraction, and general bioturbation. They were discovered on the ground surface by standard archaeological pedestrian survey.

Each of the GPR reflection profiles collected in this large area displayed very different buried materials, and as only one day of data collection was devoted to this project only. A very coarse grid of transects spaced between 25 and 50 m apart was collected. This grid of profiles did not allow good correlation of geological units between adjacent profiles, and therefore only two-dimensional images of the ground could be generated. These images were informative and important, but only allowed for the generation of working hypotheses regarding the geological and environmental history of this area.

This project was initiated knowing that it would necessarily be a preliminary study that could provide data for more complete geoarchaeological analyses in the future. However, some very interesting geological units are visible in the profiles, and a basic understanding of the geological history of this area of the floodplain was possible. This allowed for a tentative placing of the Late Paleolithic people, who exploited this coastal area, into a well-defined ancient landscape.

The GPR reflection profile that began in the middle of the dry lake bed, and continued about 93 m to the east, clearly shows Jurassic bedrock at about 7 m depth, consistent with projections of that bedrock unit from our excavations just to the west of the lake (**Figure 1.4**). Resting directly on an unconformable surface with Jurassic are Pleistocene

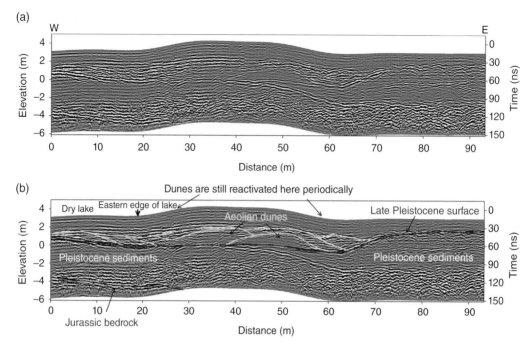

Figure 1.4 Two hundred and seventy megahertz reflection profile from western Portugal used to define subsurface sediment packages from late Pleistocene through recent times. Profile (a) is unannotated while (b) shows the inferred sediment packages, correlated to outcrops about 1 km away.

sediments of unknown age, which have the appearance in the two-dimensional GPR profiles of sand dunes. These units have been described elsewhere along the Portuguese coast as near-shore and aeolian environments (Benedetti et al. 2009), but have not been studied in this immediate area along what would have been an interior floodplain environment. On beach cliffs about 4 km to the west on the other side of the coastal hills, these Pleistocene age units date to about 62,000 years BP.

What is more important to the goals of this study is the reflection surface in the GPR profile (colored in blue in the lower annotated profile in **Figure 1.4**), which is consistent stratigraphically with a buried soil visible in outcrop about 1 km to the west. This unit was formed during a period of landscape stability during the Late Pleistocene when a well-developed soil was formed (Conyers et al. 2013). It was just below this buried soil unit, dating to about 11,500 BP, where late Paleolithic age artifacts were found in place along the edge of a small fluvial channel incised into the Jurassic bedrock. Overlying this late Pleistocene surface (**Figure 1.4**) are a sequence of sand dunes, clearly visible as a progressively thickening sand package of large-scale forest beds in individual dunes, preserved west of the present-day dry lake bed. These dunes appear to overlay the continuous late Pleistocene surface, which is likely the buried soil known from outcrops to the west. The aeolian sand units thicken to the east, toward the river, and possibly are the damming feature that created the now-dry lake, still visible on the surface.

It was initially hypothesized that this lake (**Figure 1.3**) was formed by salt collapse in the Jurassic bedrock, which has been documented elsewhere in this part of western

Portugal (Benedetti et al. 2009). The GPR profile indicates that a more likely origin for the lake is the thick aeolian sand unit that blocked water runoff derived from the high coastal bedrock ridge to the west. There is no evidence of any collapse features in the Jurassic bedrock or other units visible in the GPR profiles collected along the margin of the lake, or at least a collapse feature that would be visible in the upper 7–8 m of this sedimentary package (Figure 1.4).

This small geoarchaeological study incorporating GPR with a minor amount of geological and archaeological background demonstrates the utility of posing geological and environmental change hypotheses and then testing them with some basic interpretations using GPR images. While these results from Portugal must remain preliminary until additional three-dimensional analysis of sedimentary units can be accomplished with more tightly spaced GPR reflection profiles, some important conclusions can still be made. In this case a basic stratigraphic analysis of geological units within the floodplain was accomplished and one hypothesis about the origin of the lake was tested.

This stratigraphic study shows that sand dunes and other unknown Pleistocene sediments rest unconformably on Jurassic bedrock, which is consistent with stratigraphy visible in outcrops some kilometers to the west. While there are some interesting units in the sediments below the late Pleistocene soil, especially as visible on the eastern edge of the reflection profile (Figure 1.4), nothing is known of their precise age and origin. They are labeled "Pleistocene sediments" on the profile and have the appearance of cross-bedded sand dunes and other less reflective sedimentary units. The late Pleistocene surface lies directly on these sediments, colored in blue, which is most likely the soil unit exposed about 1 km to the west in a number of outcrops. Stratigraphically just below this soil unit late Paleolithic artifacts were discovered in place, in what was interpreted as a hunting or other short-term camp (Conyers et al. 2013; Haws et al. 2011). At the very end of the Pleistocene sand dunes then covered the soil unit, with thicker accumulation eastward toward the river. These dunes likely acted as a dam for water running off the uplifted coastal hills to the west and a lake was formed after the artifacts were deposited in the unit to the west near the base of this sand layer. It then appears that Late Paleolithic people continued to be drawn to this area after the lake formed and additional stone tools were deposited near its margin, which were incorporated into the accumulating dune units visible on the GPR profile to the east of the lake. These dunes continued to be actively deposited through much of the Holocene, and are visible as unconsolidated surface sand today (Figure 1.3). The artifacts found on the surface in this area were likely brought to the surface by plowing or possibly reactivation of the dunes over time, exposing and then reworking materials from deeper in the sedimentary sequence.

This simple example of the utility of GPR in geoarchaeology shows how a small amount of archaeological materials, when placed into a stratigraphic sequence that is generally dated, can produce working hypotheses regarding the location and nature of ancient environments and environmental change over time. When the ancient people who left these tools are then placed within that framework other conclusions can be made about those hunting and gathering people who exploited the landscape and what resources were important to them.

These very basic conclusions can be expanded in the future with additional GPR profile collection in a finer grid of data, and with coring and age dating of these buried units. In this example one GPR profile allowed for important hypotheses regarding how environments changed and evolved during about a 40,000 year time period. The basic sedimentary packages, within which the artifacts were found, were defined geophysically, and new hypotheses formulated about the environmental history of this area and people's exploitation of the ancient landscape. While much in this simple example remains speculative, it shows how the integration of information from three separate disciplines (archaeology, geology and geophysics) can yield a great deal of important data that can "drive" new ideas and hypotheses about this coastal floodplain in western Portugal and its late Paleolithic history.

References

Behrensmeyer, Anna K. & Hill, Andrew P., eds. (1998) *Fossils in the Making: Vertebrate Taphonomy and Paleoecology.* University of Chicago Press, Chicago, Illinois.

Benedetti, Michael M., Haws, Jonathan A., Funk, Caroline L., et al., (2009) Late Pleistocene raised beaches of coastal Estremadura, central Portugal. *Quaternary Science Reviews,* vol. *28,* no. 27, pp. 3428–47.

Birkeland, Peter (1999) *Soils and Geomorphology,* 3rd Edition. Oxford University Press, New York.

Bruno, David & Thomas, Julian, eds. (2008) Landscape archaeology: introduction. In: *Handbook of Landscape Archaeology*, pp. 27–43. Left Coast Press, Walnut Creek, California.

Butzer, Karl W. (1971) *Environment and Archeology: An Ecological Approach to Prehistory.* Aldine Publications Co., New York.

Campana, Stefano & Piro, Salvatore, eds. (2008) *Seeing the Unseen – Geophysics and Landscape Archaeology.* Taylor & Francis, London.

Constante, Ana, Peña-Monné, José Luis, & Muñoz, Arsenio (2010) Alluvial geoarchaeology of an ephemeral stream: implications for Holocene landscape change in the central part of the Ebro Depression, Northeast Spain. *Geoarchaeology,* vol. *25,* no. 4, pp. 475–96.

Conyers, Lawrence B. (2009) Ground-penetrating radar for landscape archaeology. In: Campana, Stefano & Salvatore Piro (eds.) *Seeing the Unseen-Geophysics and Landscape Archaeology,* pp. 245–56. CRC Press/Balkema: Taylor and Francis Group, London.

Conyers, Lawrence B. (2012) *Interpreting Ground-penetrating Radar for Archaeology.* Left Coast Press, Walnut Creek, California.

Conyers, Lawrence B. (2013) *Ground-penetrating Radar for Archaeology,* 3rd Edition. Altamira Press, Rowman and Littlefield Publishers, Lantham, Maryland.

Conyers, Lawrence B., Daniels, J. Michael, Haws, Jonathan A., & Benedetti, Michael M. (2013) An upper Palaeolithic landscape analysis of coastal Portugal using ground-penetrating radar. *Archaeological Prospection,* vol. *20,* no. 1, pp. 45–51.

Davidson, Donald A. & Shackley, Myra L., eds. (1976) *Geoarchaeology: Earth Science and the Past.* Westview Press, Boulder, Colorado.

Ferring, C. Reid (2001) Geoarchaeology in alluvial landscapes. In: Paul Goldberg, Vance T. Holliday, & C. Reid Ferring (eds.) *Earth Sciences and Archaeology,* pp. 77–106. Springer US, New York.

French, Charles (2003) *Geoarchaeology in Action: Studies in Soil Micromorphology and Landscape Evolution.* Routledge, London.

Gaffney, Chris, Gaffney, Vince, Neubauer, Wolfgang et al. (2012) The Stonehenge hidden landscapes project. *Archaeological Prospection,* vol. *19,* pp. 147–55.

Goldberg, Paul, Holliday, Vance T., & Ferring, C. Reid, eds. (2001) *Earth Sciences and Archaeology.* Kluwer Adacemic/Plenum Publishers, New York.

Haws, Jonathan A., Funk, Caroline L., Benedetti, Michael M. et al. (2011) Paleolithic landscapes and seascapes of the west coast of Portugal. In: Nuno Bicho, Jonathan A. Haws, & Loren Davis (eds.) *Trekking the Shore*, pp. 203–46. Springer US, New York.

Holliday, Vance T. (2001) Quaternary geoscience in archaeology. In: Paul Goldberg, Vance T. Holliday, & C. Reid Ferring (eds.) *Earth Sciences and Archaeology,* pp. 3–35. Springer US, New York.

Johnson, Jay K., ed. (2006) *Remote Sensing in Archaeology: An Explicitly North American Perspective.* University of Alabama Press, Tuscaloosa, Alabama.

Kvamme, Kenneth L. (2003) Geophysical surveys as landscape archaeology. *American Antiquity*, vol. *63*, no. 3, pp. 435–57.

Rapp, George & Hill, Christopher L. (2006) *Geoarchaeology: The Earth-science Approach to Archaeological Interpretation,* 2nd Edition. Yale University Press, New Haven, Connecticut.

Ricklis, Robert A. & Blum, Michael D. (1997) The geoarchaeological record of Holocene sea level change and human occupation of the Texas gulf coast. *Geoarchaeology*, vol. *12*, no. 45, pp. 287–314.

Schiffer, Michael B. (1972) Archaeological context and systemic context. *American Antiquity*, vol. *37*, no. 2, pp. 156–65.

Shackley, Myra L. (1975) *Archaeological Sediments: A Survey of Analytical Methods.* John Wiley & Sons, Inc., New York.

Stern, Nicola (2008) Stratigraphy, depositional environments, and paleolandscape reconstruction in landscape archaeology. In: David Bruno & Julian Thomas (eds.) *Handbook of Landscape Archaeology*, pp. 365–78. Left Coast Press, Walnut Creek, California.

Trinks, Immo, Johansson, Bernth, Gustafsson, Emilsson et al. (2010) Efficient large-scale archaeological prospection using a true three-dimensional ground-penetrating radar array system. *Archaeological Prospection*, vol. *17*, pp. 175–86.

Waters, Michael R. (1992) *Principles of Geoarchaeology, A North American Perspective.* The University of Arizona Press, Tucson, Arizona.

2

Basic Method and Theory of Ground-penetrating Radar

Abstract: The usual method of collecting GPR data is stacking traces, consisting of many vertically acquired reflections, along transects to produce reflection profiles, collected within programmed time windows that represent depth in the ground. If stratigraphic interfaces in the ground are complex, each reflection profile must be interpreted individually to determine the origin of reflections. The amplitude of reflections varies with sediment types, porosity, composition, and water retention. Resolution of stratigraphy is a function of the thickness of units and the wavelength of the propagating radar waves in the ground. High frequency antennae produce short wavelength waves, which can resolve thinner bedding and smaller objects, but often only at shallow depths. Lower frequency antennae only resolve thicker beds but often at greater depths. After reflection origins have been determined and can be understood in two-dimensional profiles, amplitude slice-maps representing the reflections along planar surfaces in map view can be produced to understand the aerial distribution of reflections in three dimensions.

Keywords: reflection profiles, bedding resolution, three-dimensional maps, frequencies, depth of penetration, reflection origins

Introduction

Ground-penetrating radar data are acquired by reflecting pulses of radar energy on a surface antenna, which generates waves of various wavelengths that propagate outward. They spread into the ground in a cone as the waves propagate downward. As these waves move downward they can be reflected from buried objects, features or bedding planes (**Figure 1.2**). The reflected waves then travel back to the ground surface and are detected and recorded at a receiving antenna that is paired with the transmitting antenna. The two-way travel times of the waves into the ground to the reflection surface and then back to the receiving antenna are recorded in nanoseconds. As the radar waves propagate through various materials in the ground their velocity will change, depending on the physical and chemical properties of the material through which they are traveling (Conyers 2013, p. 24). At contacts between different materials in the ground the waves' propagating velocity can change and when this occurs a reflected wave is generated. Some reflected waves will then travel back to the ground surface while the remaining energy continues to propagate deeper and can be reflected

Ground-penetrating Radar for Geoarchaeology, First Edition. Lawrence B. Conyers.
© 2016 John Wiley & Sons, Ltd. Published 2016 by John Wiley & Sons, Ltd.

again from additional deeper interfaces, until all the energy finally dissipates with depth due to absorption by the ground and spherical spreading with depth. Only the reflected energy that travels back to the surface antenna is recorded and visible for interpretation.

The velocity of radar energy in the ground can be calculated and reflected radar wave travel times converted to distance (or depth in the ground). It is this ability to determine depth that makes GPR capable of producing a three-dimensional dataset. There are many ways to calculate velocity (Conyers 2013, p. 107; Conyers and Lucius 1996), all of which are estimates of wave propagation speed through packages of sediments and soils. Velocity of propagating waves can vary considerably with depth, usually decreasing as water saturation increases with depth of burial, and also laterally because of a variety of other changes in ground composition.

In most GPR datasets radar antennae are moved along the ground in transects and two-dimensional profiles of a large number of reflections at various depths are created to produce many reflection profiles for interpretation (**Figures 1.2** and **2.1**). When data are acquired in a closely spaced series of antenna transects within a grid, reflections from adjoining profiles can be resampled, compared, gridded, and then processed into amplitude maps (**Figure 2.1**). These images produce an accurate three-dimensional

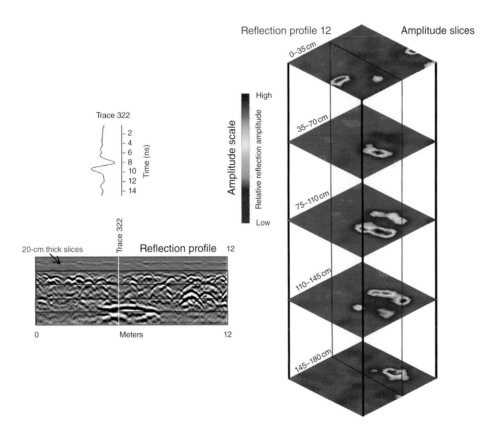

Figure 2.1 The production of GPR images, beginning with individual traces from one location on the ground, stacked together to produce reflection profiles, with an additional product being amplitude slice-maps of resampled reflection trace amplitudes in individual maps from programmed depths in the ground.

Figure 2.2 Some GPR antennae commonly used in geoarchaeological investigations with the 270 MHz energy traveling deepest in the ground (up to 6 m), with the least feature and stratigraphy definition and the 900 MHz with shallow penetration of 1–2 m maximum and higher resolution. The 400 MHz antenna is most useful for medium resolution at depths from 1 to 4 m.

picture of reflection surfaces (Conyers 2013, p. 166) indicating the location of features spatially (in x and y dimensions) and with depth (z). An interpretation of the reflections in the ground can then be accomplished using these standard GPR images, which are reflection profiles and amplitude maps. Other images can also be produced such as isosurface renderings and videos of layers and features in the ground, but for most geoarchaeological studies these are not as useful in complex layered ground.

For most geoarchaeological studies reflection profiles are the most useful GPR tool, as they illustrate stratigraphic units that can be most easily interpreted for geological purposes. Amplitude slice-maps can also be useful interpretive tools sometimes, but as they tend to cross bedding planes especially in complexly layered ground, they tend to be somewhat cluttered with anomalous reflections that are more difficult to interpret. Usually for stratigraphic interpretations each of the profiles must be interpreted individually, which can be time consuming, but much more exact than the "batch processing" used to produce amplitude slices (**Figure 2.1**).

Production of Reflections, Depth of Penetration, and Resolution

The buried discontinuities where reflections occur are usually created by changes in electrical properties of the sediment or soil, lithological changes, and differences in bulk density at stratigraphic interfaces. These measurable (and sometimes visible) differences in materials in the ground create water saturation differences within the stratigraphic layers, which is what usually produces the velocity changes that generate wave reflections (Conyers 2012, p. 34; 2013, p. 47).

The depth to which radar energy can penetrate and the resolution that can be expected in the subsurface are partially controlled by the frequency of the radar energy transmitted. The propagating radar energy frequency controls both the wavelength of the transmitted waves and their spreading, and the amount of weakening, or **attenuation**, of those waves in the ground. Standard GPR antennae used in geoarchaeology propagate radar energy that varies in bandwidth from about 10 megahertz (MHz) to 900 MHz (Figure 2.2). Some very rare applications in small-scale, very high-resolution studies employ frequencies higher than 900 MHz. Antennae usually come in standard frequencies, with each antenna having one central frequency, but actually producing radar energy that ranges a good deal around that mean. Each GPR system manufacturer produces different frequency antennae, none of which are interchangeable with the other's systems, and each of which has its own interesting and sometimes infuriating pros and cons.

The 400 MHz antenna (Figure 2.2) is usually capable of transmitting radar energy to about 2–3 m in depth with resolution of bedding planes of features down to about 20 cm or so. The 270 MHz antennae are often effective in propagating energy to greater than 2 m in the ground but the longer wavelengths generated from them have less subsurface resolution. In addition, the 270 MHz frequency energy spreads out more from the transmitting antenna, and as the waves move into the ground they reflect from a larger volume of material. In addition, surface walls, trees, livestock, and people will all produce reflections from the lower frequency antennae due to the greater energy spreading, which can be frustrating when trying to differentiate them from the reflections generated within the ground.

The high-frequency antennae, such as the 900 MHz, are quite good at shallow mapping within a meter or so of the ground surface and have very good feature resolution to about 10 cm or so. However, if high frequency antennae are used in even moderately electrically conductive ground their transmitted energy can be attenuated within 30–40 cm of the ground surface. The downside to using antennae in this frequency range is that they also record within the band width of personal communication devices, and there can be more background noise from cell phone (mobile phones) and various other radio transmissions, along with the desired reflected waves from within the ground (Conyers 2013, p. 80).

Very low frequency antennae in the 10–100 MHz range are all unshielded, meaning they have no ability to block radar wave transmission outward and can record reflections generated from all directions, and therefore can be quite "noisy" with the recording of many anomalous reflections. They are also only useful in very electrically resistive ground where depth penetration is important (3–15 m depth) and the resolvable stratigraphic interfaces and features tend to be measured in meters rather than centimeters.

Data Collection and Recording

The most efficient collection method for subsurface GPR mapping is to establish a grid across a survey area prior to acquiring reflection data. Usually rectangular grids consisting of many reflection profiles are collected within that grid (Figure 2.1) often with a transect spacing of 1 m or less. This will produce a very precise stratigraphic picture of the ground, but can take a long time to collect and interpret. Sometimes for large scale geoarchaeological studies, a much wider transect spacing will suffice, especially if only the larger scale geological units are of interest. The higher frequency antennae that focus energy more in the ground and have a much higher bedding plane and feature resolution, usually necessitate closer transect spacing than those of lower frequency, if complete coverage is desired.

If only basic stratigraphy is necessary, GPR transects need not be located in a rectilinear grid, but placed on the ground surface where antennae can be most easily moved or where they will cross geological layers or buried units of interest in an appropriate direction. Profiles of this sort can be placed over topographically complex ground, and even around surface barriers or obstacles, as long as their locations are surveyed in and elevations changes are determined by GPS or other surveying methods.

If profiles are collected in rectilinear grids a Cartesian coordinate system can be used to place reflections into space, which then can be readily processed using most standard available mapping and gridding software. A few systems are in use that can place randomly collected reflections into space using GPS-collected spatial data. It will not be long before spatially random data collection can be processed into a three-dimensional volume, and profiles and amplitude maps can then be produced from a package of reflections in any dimension or orientation chosen by the interpreter. This technology is only a year or so away from wide application in GPR and has been commonly used in the petroleum industry for seismic wave reflection interpretation for a number of years, with very dense and complex three-dimensional datasets.

During GPR data collection the two-way travel time and the amplitude and wavelength of the reflected radar waves derived from within the ground are amplified, processed and recorded for immediate viewing and later post-acquisition processing and display. At any one specific location along an antenna transect the waves that are recorded and stacked vertically from many depths in the ground is called a reflection trace (**Figure 2.1**). When many traces are recorded sequentially as antennae are pulled along the ground surface in a transect (with perhaps up to 50 traces recorded every meter depending on the resolution desired) a reflection profile is produced (**Figure 2.1**). Distance along each antenna transect is recorded for accurate placement of all reflections in space using a survey wheel (**Figure 1.1**), manual **distance markers** or GPS receivers connected to the antennae.

Radar energy becomes both dispersed and waves become attenuated as they radiate from the surface antenna into the ground (**Figure 1.2**). When portions of the original transmitted waves are reflected from buried interfaces back toward the surface they will suffer additional attenuation by the material through which they pass, before finally being recorded at the surface. To be detected as reflections, important subsurface interfaces must not only have sufficient contrast at their boundary to abruptly slow or speed up wave propagation and produce a reflection, but also must be located at a shallow enough depth where sufficient radar energy is still available to be reflected. Attenuation occurs when the radar energy travels to increasing depths and the waves becomes weaker as they spread out over more surface area and are progressively absorbed (conducted away) by the weak electrical properties of most ground. For every site the maximum depth of radar wave penetration will vary with the ground chemistry (with the clay conductivity and water content being the most important variable) and many other geologic and moisture conditions (Conyers 2013, p. 62).

Production and Processing of Reflection Profiles

To create a vertical display of the subsurface reflections, all recorded reflection traces are displayed in a format where the two-way travel time, or approximate depth when converted from time using velocities, is plotted on the vertical axis with the surface location on the horizontal axis (**Figure 2.1**). Most visualization software allows the horizontal and vertical scales to be adjusted in order to display reflections with any exaggeration necessary to allow reflections to be interpreted. When there is significant elevation variation along a survey transect, topographic corrections should also be made that will adjust the recorded reflections for surface variations and display those reflections in the ground in more realistic profiles.

While there are a number of data filtering and adjustment processes that are commonly applied to reflection data (Conyers 2013, p. 129), only a few are used almost all the time prior to interpretation. The most common data processing method employed by all GPR software is range gaining. Due to the conical downward spreading of the transmitted radar waves and the attenuation of radar energy as it passes through the ground, later reflection arrivals recorded from interfaces deeper in the ground will almost always have lower amplitudes than earlier arrivals. To recover these lower-amplitude waves, gain control (range gaining) is then applied to all reflection traces in a profile during either acquisition or post-acquisition processing (Conyers 2013, p. 99; Jol and Bristow 2003; Neal 2004). This computer function will amplify those waves received from deeper in the ground so they become visible. If reflection data are highly attenuated with depth in the ground, often no reflections will be received from below a certain depth and the only energy being recorded is external or system noise (Conyers 2013, p. 33). That attenuation depth will be apparent in reflection profiles as the location where no coherent reflections are being recorded and displayed.

Other common processing steps, which are suggested by many in the GPR community (Woodward et al. 2003), include vertical filters, also called **band-pass filters**, employed to remove anomalously high- and low-frequency noise during data recording and also during data analysis and interpretation (Conyers 2013, p. 134; Neal 2004). Terms for this **filtering** are high-pass and low-pass with the high-pass filter removing low-frequency waves often generated from "system noise" inherent to each particular radar device and low-pass filters removing high frequency noise from of various sorts of external radio transmissions. In some GPR systems these filters can be applied during data acquisition, and in others they must be applied during data processing prior to displaying reflection profiles for interpretation.

Another data processing tool commonly used for all image enhancements is **background removal**, which creates a composite trace of waves that were recorded in all or some number of sequential traces in a profile and then removing that average trace from each trace within the profile (Neal 2004). The method relies on the fact that reflections recorded at the same time in a profile, which exhibit the same wave "signature" within a running series of traces, will have likely been generated by background noise that obscures reflected waves generated from within the ground. That background noise can then be removed from all the reflections in a profile, retaining and displaying only those that were obtained from within the ground and were likely to have been recorded at different times and with different amplitudes. In most cases this simple procedure creates a much "cleaner" reflection profile where only those reflections of interest are displayed (Conyers 2013, p. 134). However there is a somewhat low risk that perfectly horizontal reflections from some interface in the ground could be removed by this procedure. Usually once background waves are removed, profiles must again be re-gained in order to visualize the remnant reflections of interest in a profile.

More Advanced Data Processing Steps

Other than the basic GPR data processing discussed above, there are becoming increasingly more filtering and image generation steps that have become common during interpretation. These are data manipulation steps that must be used only for specific purposes, and should usually not be part of initial interpretation methods as they can produce artifacts of unknown origin and confuse interpretation (Conyers 2012, p. 42; Woodward et al. 2003). Trace stacking will help to overcome problems with antenna coupling resulting from moving them along un-even ground and also remove other noisy or distorted traces in a profile. It produces a more "average" reflection profile, removing small and localized reflections generated from point sources and retaining extensive planar reflections. The risk is that by removing all reflection hyperbolas, important smaller reflection features will also disappear. Hyperbolas are important reflections as they are generated from "point sources" in the ground such as rocks, tops of walls, or anything that is an isolated feature (Conyers 2013, p. 57).

Another data processing step that is commonly used prior to the production of amplitude maps is **migration** (Conyers 2013, p. 130). Migration is a basic data processing step that can move radar reflections to a more accurate location in a reflection profile (Conyers 2013, p. 131). A distortion of reflections positions in space can be caused by radar waves that have moved into the ground in a conical transmission pattern and are then recorded in a location not directly below the surface antennae. The most common of these distorted reflections are hyperbolas generated from individual "point sources." The migration processing step can enhance the amplitudes generated at the apex of reflection hyperbolas, while removing their axes. To perform this step, velocity estimates of the ground must be determined so that hyperbolas can be affectively removed, as their geometry is a function of velocity. This is always difficult, as in most ground conditions, velocity changes with depth (usually slowing) and also varies laterally. A velocity that might be used for all hyperbolic reflection migrations can over or under-migrate many, sometimes producing very blurred or distorted reflection profiles. Migration can be a particularly beneficial processing step prior to producing amplitude slice-maps, as migrated profiles will be much more "crisp" and less distorted by hyperbolas. Migration is also a processing step that can be used to correct steeply dipping layers that are distorted by radar wave movement in a non-vertical path from the surface antenna (Jol and Bristow 2003, p. 21).

Another useful processing step, which can be applied to data after collection, is post-acquisition frequency filtering (Conyers 2013, p. 129) that can produce profiles that display only reflections of certain desired wavelengths. This is possible because most GPR antennae are "wide-band" transmitters and receivers, which produce and record a range of wave frequencies around a mean. Therefore, when an antenna is defined as having a frequency of 400 MHz, this really means that it is generating energy on either side of an estimated "center frequency" and there are really many waves produced and collected of many frequencies. The rule of thumb is that in common GPR broad-band antennae the frequencies vary between ½ and 2 times the center frequency, which for the 400 MHz antennae would mean they are generating waves that move into the ground between about 200 and 800 MHz (Conyers 2013, p. 64). While most of the energy produced varies only a small amount from the mean, some of the other frequency waves that are generated will still move into the ground to also be being reflected and recorded back at the surface. The frequencies of waves recorded are a function of a wide range of chemical and physical variables within the materials through which the energy passes (Annan 2008). As a result, what is displayed in a reflection profile, or what is re-sampled from profiles to produce amplitude maps, are averages of all the waves recorded in many thousands of individual reflection traces.

To frequency-filter GPR data, all traces in a profile must be re-sampled using software that can remove some frequencies while retaining others to produce a whole new set of reflection profiles with different traces. The higher frequency energy is shorter in wavelength, and therefore will usually reflect from smaller buried features or objects in the ground creating better spatial definition of reflections, but from a shallower depth. Lower, longer wavelength, waves will theoretically travel deeper in the ground, but are only reflected from the larger objects or thicker layers and processing steps that will display only those units in profiles can be sometimes beneficial. This post-acquisition processing step can be useful, if is found during the interpretation process that smaller (or larger buried features) need to be better resolved, and only one frequency of antenna was used to collect the original reflection data.

Depending on the questions about buried features and units of interest asked during interpretation, some of these post-acquisition processed images can be generated to address certain problems. They should not be employed as a first step until the raw data are processed and the resulting displays are interpreted in the more standard ways outlined above. At that time a second or third version of the same data can be produced and compared to previous images in a way that can help define stratigraphy and buried features of interest. Data processing and interpretations of this sort that employ many processing steps can take some time, as all these data manipulation stages must be performed in a knowledgeable and deliberate way, but the outcomes can often be informative.

Interpretation of GPR Reflections in Profiles

Prior to making a GPR interpretation based on a visual analysis of the standard processed reflection profiles an understanding of what causes radar reflections in the ground is necessary. This is not necessarily a simple process as there are many variables that affect radar wave propagation and reflections, some of which are a function of the GPR equipment that is employed. Other variables include ground conditions, collection settings and background noise, and all must be taken into account. Most important to all GPR image interpretation is a basic analysis of what produces reflections of radar waves that are propagating and then reflecting as they move through layers in the ground (Conyers 2012, p. 32; 2013, p. 47).

In the usual way that GPR is used for purely archaeological feature mapping practitioners sometimes refer to interesting radar reflections that are visible in profiles, and in amplitude slice maps, as "anomalies" (Conyers 2012, p. 48). This description of radar energy reflections is vague and mostly un-interesting, as very few reflections are anomalous, and all have been produced for a reason that can, and usually must, be understood. Geological users of GPR almost never fall into this inexact interpretative description as all stratigraphic interfaces of interest could be termed "anomalies" in layered ground. In geoarchaeology there is no need to rely on these types of vague interpretative descriptions and instead important stratigraphic units, soils and other interfaces must be described first in geophysical terms and then interpreted geologically and archaeologically based on what created them in the ground.

The first step in any interpretative process with GPR profiles is to describe radar reflections in geophysical terms (Conyers 2012, p. 129). For instance, a common reflection feature visible in many GPR profiles might be described simply as a "high amplitude point source reflection hyperbola with an apex and axes of a certain geometry and amplitude." This type of "point source" reflection might have been generated from a buried rock, the top of an archaeological wall or some other aerially-restricted feature. Another common feature in profiles could be described as a "high amplitude undulating planar reflection" located at a specific depth, which might be a stratigraphic interface between differing sediment types (Conyers 2013, p. 62). These types of basic geophysical descriptions are the first step toward defining reflections in a way that others with some geophysical understanding can begin to visualize and understand (Conyers 2012, p. 40). Only then can these reflection features be interpreted as being produced from specific buried layers or objects of various composition and geometry. The third step is to relate these stratigraphic units to sedimentary units in specific depositional environments.

The last step in using GPR for geoarchaeology is the most important, which is to understand the origin of geological units visible with GPR, how associated cultural features were constructed, used, modified, abandoned and then encased and covered by soils and sediments over time. It is common to be unaware of the larger geological context of an archaeological site at the time of GPR data collection and often initial interpretations must be tentative. This is especially the case if GPR data are being collected well before extensive excavations have taken place or if there are no nearby exposures of the materials of interest in the ground.

In order begin to advance initial geophysical descriptions of GPR data, which can later lead to geological and archaeological interpretations, it is first important is to take into consideration what causes radar wave to be reflected from units in the ground (Annan 2008; Conyers 2013, p. 59; Neal 2004). All radar wave reflections moving through the ground are generated at interfaces where materials with different physical and chemical properties are in contact along a boundary. If the propagating radar waves change their velocity at those interfaces abruptly, radar energy will be reflected (Conyers 2013, p. 92). An example of a composition change that might affect velocity in this way would be where a clay layer overlays a sandy organic-rich soil, and where these two units are encased by more homogeneous material. The contact of the base of the clay with the sandy soil will generate a reflection as the radar waves propagating from the ground surface antenna as waves will slow abruptly as they intersect the clay layer, and then increase again as they pass into the underlying sand. The abrupt velocity changes would theoretically create reflected waves (Conyers 2013, p. 47) that would then travel back to the surface to be recorded. Those propagating and reflected waves are visible in GPR images as sine-shaped waves with both positive and negative deflections, recorded digitally as reflection traces (**Figure 2.1**).

In contrast with distinctly layered ground, a gradational change in materials over some distance would not produce a reflection as there would not be any abrupt change in radar velocity and no reflected waves would be generated. This kind of gradational change in geological contexts might occur if sediment in one layer changes from silt to sand over a distance of a meter or so. In general, the greater the change in velocity across a boundary, the greater the amplitude of the wave that is reflected back to the surface and recorded (Conyers 2012, p. 36).

When viewed in GPR reflection profiles, the geometry and amplitude of reflections visible is an excellent way to move from basic geophysical descriptions to interpretations of the genesis of layers in the ground. These types of interpretations, in various depositional settings, are based on bed geometry, sedimentary structures and contacts between units. A description and analysis of these geological features, which can be related to the archaeological record, is what makes up the majority of this book. However, before this can be done, it is important to understand first that it is not the sediment variations *per se* in the ground that are producing the reflection, but instead the ability of those materials to retain and distribute water (Conyers 2012, p. 34). For instance, a final interpretation of a GPR profile might describe a contact between a sand and clay unit. But what is really being imaged with GPR is the ability of those two units to retain and distribute water within their void space and therefore produce a radar wave reflection (Van Dam et al. 2003). This is because those two sedimentary units have very different amounts of water within them, and it is their moisture percentages that produce velocity changes as radar waves cross the interface between the two, which produces the reflections.

An understanding of what causes radar reflections and why the dominant variable in the reflectivity of various materials is water saturation is described elsewhere (Conyers 2012, p. 34). This phenomenon is generally visible when comparing a GPR reflection profile to the actual stratigraphy and archaeological features visible in the correlative outcrop as shown in **Figure 1.2**. In this geoarchaeological study a burned clay floor of an ancient pit structure is buried by aeolian dunes along the Oregon coast. The house floor is visible in the GPR profile as a high amplitude planar reflection, as it is composed of imported clay, which was compacted and partially burned by fires. This unit therefore retains much more water (or water is pooled on its upper surface) than the sand units above and below, and therefore generates distinct radar wave reflections at the contact of the clay floor with the overlying drier sand. A close comparison of the visible structures in this sedimentary package to those that produce reflections in the GPR profile is informative, as there are a number of planar reflections visible in the reflection profile that are not visible in the outcrop. There must be changes in porosity and water retention within some of these sand units that are not visible in the outcrop. Conversely there are some visible changes in the sediments that are not related to water retention and therefore those units do not generate radar reflections.

To demonstrate this water retention concept in the generation of reflected waves, a simple model is illustrative (**Figure 2.3**). In this model a dry sand layer overlies a clay unit, with a sharp interface between them. Using **relative dielectric permittivity (RDP)** values for these materials, which is a proxy measurement for radar wave velocity (Conyers 2013, p. 48), a simple set of calculations can allow for a basic understanding of how water differences relate to reflection generation and amplitude. The model demonstrates how no reflected wave will be produced at the contact between these two units when both are totally dry (**Figure 2.3a**). The RDP values between these two clastic constituents are almost the same, and therefore there was no radar wave velocity change when the interface was crossed by propagating radar energy and no reflected wave is generated. These measurements of RDP have been validated using numerous sediment and soil samples that have been dried in an oven in the laboratory and then velocity through them measured (Conyers 2013, p. 123). In these models the only way to obtain a greater contrast in RDP between these two units, is to add water, which is retained in the available void space. Fresh water has a RDP of 80, in contrast to almost all dry sediments that have an RDP between three and five (Conyers 2012, p. 34). Therefore the addition of only a small amount of water to the sediments, which is retained in the void space, will produce a material with an overall much higher RDP. The greater the difference in the amount of retained water between the units, the higher the amplitude of the reflected wave generated at the interface. When the clay unit (with a 40% porosity)

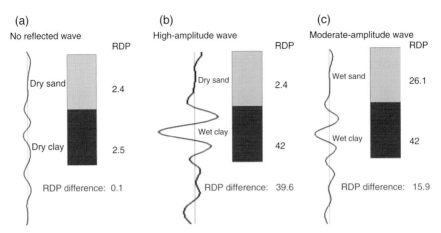

Figure 2.3 Models showing the differences in reflectivity from one interface between a clay and sand unit under different water conditions. When both are dry (a), there is no reflected wave at the interface, but when water is added to the system and retained only in the impermeable clay (b), a high-amplitude reflection is produced. When both units are wet (c) the difference in retained water in pore spaces between the two materials still produces a velocity change and therefore a reflection, but one with lower amplitude. From Conyers (2012). © Left Coast Press, Inc.

is water saturated and the overlying sand is dry, the amplitude of the reflected wave generated at their interface is high as the difference in RDP between these two units is almost 40 (**Figure 2.3b**). When both units are wet, there is still a difference in RDP as these two sediments have a very different porosity, and therefore the amount of water retained in their void spaces space is different, and a reflected wave is still generated but with a lower amplitude.

The purpose of these simple models is to demonstrate differences in RDP (that create a velocity contrast) is primarily a function of the amount of water retained in the pore spaces of bounding units. While the reflections produced from the interface between two sedimentary units would be interpreted as sand and clay beds when visible in GPR profiles, what is really being measured with GPR are each unit's ability to retain water. This can often lead to some confusion during interpretation, as not all sedimentary units produce reflections and there can also be reflections created from sediments and soils with different porosity and water saturation that are not necessarily visible to the human eye when exposed in outcrops or excavations. This is part of the inherent imprecision in the GPR method, but usually these porosity changes and water retention differences are directly correlative with meaningful lithological changes that are important in geoarchaeological studies.

Resolution of Stratigraphic Units

The ability to resolve stratigraphy and associated buried archaeological features or other horizons with GPR is largely a function of the wavelength of radar energy reaching them at the depth they are buried. There are many "rules of thumb" regarding the minimum object size that can be resolved at various frequencies, which vary between 50 and 40% of the wavelength that encounters them (Jol and Bristow 2003, p. 11; Orlando 2007; Yilmaz 2001). As radar energy moves from the transmitting antenna into the ground it changes it wavelength (and therefore frequency) as velocity decreases in sediments and

Table 2.1 Length (in meters) of radar waves in media of a given RDP and frequency.

RDP	Frequency (MHz)									
	100	200	300	400	500	600	700	800	900	1000
1	2.998	1.499	0.999	0.750	0.600	0.500	0.428	0.375	0.333	0.300
2	2.120	1.060	0.707	0.530	0.424	0.353	0.303	0.265	0.236	0.212
3	1.731	0.865	0.577	0.433	0.346	0.288	0.247	0.216	0.192	0.173
4	1.499	0.750	0.500	0.375	0.300	0.250	0.214	0.187	0.167	0.150
5	1.341	0.670	0.447	0.335	0.268	0.223	0.192	0.168	0.149	0.134
6	1.224	0.612	0.408	0.306	0.245	0.204	0.175	0.153	0.136	0.122
7	1.133	0.567	0.378	0.283	0.227	0.189	0.162	0.142	0.126	0.113
8	1.060	0.530	0.353	0.265	0.212	0.177	0.151	0.132	0.118	0.106
9	0.999	0.500	0.333	0.250	0.200	0.167	0.143	0.125	0.111	0.100
10	0.948	0.474	0.316	0.237	0.190	0.158	0.135	0.119	0.105	0.095
11	0.904	0.452	0.301	0.226	0.181	0.151	0.129	0.113	0.100	0.090
12	0.865	0.433	0.288	0.216	0.173	0.144	0.124	0.108	0.096	0.087
13	0.831	0.416	0.277	0.208	0.166	0.139	0.119	0.104	0.092	0.083
14	0.801	0.401	0.267	0.200	0.160	0.134	0.114	0.100	0.089	0.080
15	0.774	0.387	0.258	0.194	0.155	0.129	0.111	0.097	0.086	0.077
16	0.750	0.375	0.250	0.187	0.150	0.125	0.107	0.094	0.083	0.075
17	0.727	0.364	0.242	0.182	0.145	0.121	0.104	0.091	0.081	0.073
18	0.707	0.353	0.236	0.177	0.141	0.118	0.101	0.088	0.079	0.071
19	0.688	0.344	0.229	0.172	0.138	0.115	0.098	0.086	0.076	0.069
20	0.670	0.335	0.223	0.168	0.134	0.112	0.096	0.084	0.074	0.067
30	0.547	0.274	0.182	0.137	0.109	0.091	0.078	0.068	0.061	0.055
40	0.474	0.237	0.158	0.119	0.095	0.079	0.068	0.059	0.053	0.047
50	0.424	0.212	0.141	0.106	0.085	0.071	0.061	0.053	0.047	0.042
60	0.387	0.194	0.129	0.097	0.077	0.065	0.055	0.048	0.043	0.039
70	0.358	0.179	0.119	0.090	0.072	0.060	0.051	0.045	0.040	0.036
80	0.335	0.168	0.112	0.084	0.067	0.056	0.048	0.042	0.037	0.034

especially moist or water saturated sediments. This velocity impendence shortens the wavelength of propagating waves (Conyers 2013, p. 64). For instance, a 400 MHz center-frequency antenna will generate propagating energy with a wavelength of 75 cm in air as it leaves the surface antenna, which will be "downloaded" to about 30 cm (Table 2.1) as it moves through material with an RDP of 6 (consistent with a sand unit that has low retained moisture). As these waves move deeper into the ground they usually encounter material of higher RDP, which occurs as water saturation increases, and wavelengths will continue to shorten with depth. In complexly layered ground with layers of different RDP, it is very difficult to predict wavelength at any given depth, so it is usually sufficient to be aware that wavelengths of propagating waves change, usually decreasing, with progressively deeper movement into the ground.

If very different velocity materials are layered in the ground, the variations in wavelength between units will be much too complicated to predict or measure. However, in ground with a RDP of 6, and 400 MHz radar waves downloading to 30 cm in the ground, objects with a diameter of about 12 cm (40% of 30 cm) will generate hyperbolic reflections that are visible in reflection profiles.

This general understanding of wavelength and resolution is important because it is necessary to employ the correct antenna frequency for the resolution of buried layers and objects desired and also to ensure energy propagation deep enough to reflect from these materials of interest. To distinguish radar reflections generated from two parallel buried layers (the top and base of a stratigraphic unit), the two interfaces must be separated by at least one wavelength of the energy that is encountering them (Baker et al. 2007; Conyers 2013, p. 67; Jol and Bristow 2003). If the two reflections generated from the two interfaces are not separated by that distance, then the resulting reflected waves generated from both can be destroyed or unrecognizable due to constructive and destructive interference as the two waves move simultaneously back to the surface to be recorded.

As an example of this important stratigraphic resolution concept in GPR for geoarchaeology, a layer of sandy silt with an RDP of 10 was cored and found to thin from 18 to 7 cm over a distance of about 2 m (**Figure 2.4**). A 900 MHz antenna with a propagating wavelength of 33 cm in air (Table 2.1) was used to collect a reflection profile across this buried layer of known thickness. The 900 MHz propagating radar waves downloaded to about 10 cm as they moved through this ground (Table 2.1), and reflected waves were produced and recorded from both the top and bottom of the 18 cm thick layer. They were recorded as two distinct waves because they are separated by a

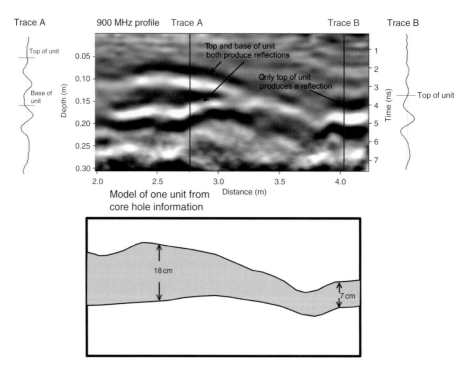

Figure 2.4 Resolution of stratigraphic interfaces is a function of radar wavelength and the distance between two interfaces. In this profile created with 900 MHz antenna reflections (upper image) the upper and lower interfaces are resolved, as each produces distinct reflections as can be seen in Trace A. When the distance between the interfaces becomes shorter there is constructive and destructive interference between the waves generated and the lower interface becomes invisible in Trace B with the lower boundary invisible in the reflection profile.

distance greater than the propagating radar wavelength in that material. However, when the transmitting antenna moved over the buried unit where it thinned to 7 cm, the propagating waves that encountered this unit were still reflected from the top and base but the later wave arrival from the bottom interface was overwhelmed by the one generated from the top, and only the upper reflection is visible in the reflection profile (Figure 2.4).

In this simple one unit example, and in most layered ground, reflections from buried layers that are thick enough will be generated from the top or base of units, but their visibility is dependent on the thickness of the layers and the wavelength of energy in the ground. Interfaces that might be separated by less than one wavelength would remain invisible in GPR reflection profiles. Theoretically, there are equations and modeling on the resolution of horizontal strata that suggest there can still be resolution of layers that are as thin as ¼ of the wavelength of energy moving in the ground (Hollender and Tillard 1998). This theoretical resolution is calculated using coefficients of reflection and the generation of constructive and destructive interference of waves reflected from parallel layers. These calculations and predictive modeling using formulae for theoretical conditions does not appear to hold up in real-world conditions such as the example provided in Figure 2.4. That example, and many other field tests (Conyers 2013, p. 62) indicates resolution is only viable in layers separated by one wavelength of the propagating radar waves within the material of a given RDP.

If higher resolution of thinner layers were important then an antenna that generates higher frequency, shorter wavelength, waves would perhaps be preferable. The trade-off is that the higher frequency energy will usually travel to shallower depths (Conyers 2013, p. 42). If reflection data from only one antenna are available and greater stratigraphic resolution is needed, frequency filters can potentially be applied to display only the higher frequency, shorter wavelength reflections.

Not only is the transmitted and propagating radar wavelength important for stratigraphic resolution, but as radar energy moves into the ground it tends to spread out within a cone from the surface antenna (Conyers 2013, p. 71). This spreading of energy with depth generates reflections from much more surface area along an interface than just a small area directly below the antenna. The greater the spreading the more averaging of waves that return to the surface antenna, and the less distinct the resulting reflections become when recorded and displayed in a profile. Lower frequency, longer wavelength, waves will spread out more with depth than those of shorter wavelength (which are more focused when leaving the antenna) and therefore produce reflections from a horizon that are more "blurry" (Figure 2.5). This averaging of many reflections from a buried surface with lower frequency waves lowers resolution of buried layers but the waves will often propagate deeper in the ground (Conyers 2013, p. 62).

An example of resolution, depth of penetration and antenna frequency, was tested at an interface in a simple stratigraphic section where the base of a cobble-filled channel rests on a Pleistocene-age sand unit (Figure 2.5). Two different frequency antennae (270 and 400 MHz) were used to test resolution of both individual objects (the cobbles) and the planar stratigraphic interface between the two layers. The two profiles were not collected along exactly the same transect, as the larger 270 MHz antennae could not be placed as close to the cliff face as the 400 MHz antennae, but the stratigraphy displayed in both is essentially the same. The interface between these two sedimentary units produces a distinct planar reflection of high amplitude using both frequencies, probably because the lower sandy-silt layer retains more water and there is a velocity contrast at the contact (Figure 2.6).

Figure 2.5 The interface tested with two different reflection profiles using different frequency antennae. The stratigraphic interface is at the boundary between an upper cobble and gravel unit and an underlying silty sand in southern Arizona. The profiles collected at this exposure are displayed in Figure 2.6.

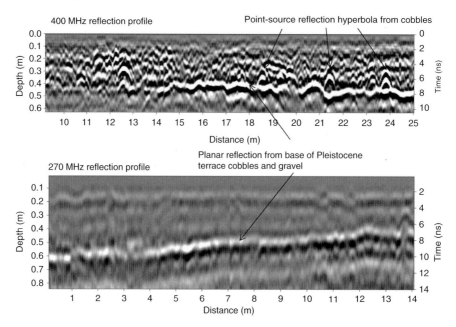

Figure 2.6 The 400 MHz reflection profile is capable of resolving most of the larger clasts in the cobble unit shown in Figure 2.5 and the interface between the two units. The 270 MHz waves passed by the cobbles with no significant reflections generated, and only the interface between the two units was observable in the reflection profile.

In the 400 MHz reflection profile the larger cobbles generated distinct reflection hyperbolas and the contact between the two units shows that this bounding surface undulates laterally (Figure 2.6). The 270 MHz data reflection profile displays the basal contact of the cobble layer with the Pleistocene sand as a much less undulating surface than the 400 MHz, with only a few amplitude changes along the interface. The variations on the planar surface in the 270 MHz reflection profile are likely a result of cobbles resting on the surface and diffracting many radar waves that were reflected away from the surface antenna.

The difference in the resolution of the stratigraphic horizon in these two reflection profiles (**Figure 2.6**) with differing antenna frequencies is illustrative. The greater averaging of reflections from the planar surface in the 270 MHz profile is primarily a function of greater spreading of energy but also a general lower resolution with the longer wavelength propagating energy. The 400 MHz energy is much more focused, and those shorter wavelength reflections generated from the stratigraphic boundary are averaged over a smaller area, and more of the undulations from this buried interface are visible in the reflection profile.

There is also a difference in the way these two profiles display different frequency reflected waves from the individual cobbles (**Figure 2.6**). The clasts in the upper fluvial unit should be good radar reflectors, but almost no reflection hyperbolas were generated from them in the 270 MHz profile. This is because the longer wavelength energy from this antenna (about 60 cm in this medium with a RDP of between 3 and 4: Table 2.1) is too long to produce reflections from objects less than about 30 cm in dimension. The "rule of thumb" for resolution of **point-source** objects is different than that used for the resolution of sediment layers discussed above (Conyers 2013, p. 69; Jol and Bristow 2003). Usually individual objects can be resolved if they are of more than about 40% of the wavelength of radar energy passing through the material. In the 270 MHz profile, with a 60 cm wavelength in this sediment, only objects larger in diameter than about 25 cm would produce hyperbolic reflections. This lower frequency antenna would therefore not be capable of generating reflections from objects of this size. If it was important to produce images of individual cobbles in the upper layer, the 400 MHz antennae would be preferable, as the wavelength of 400 MHz energy in this ground with an RDP of 3–4 is about 33 cm (Table 2.1), making objects larger than about 13 cm (40% of 33 cm) visible as hyperbolic reflections.

Another interesting comparison of the two frequency antennae in this simple two-layer example (**Figure 2.6**) is that both profiles exhibit energy attenuation at almost the same depth in the ground. This is because below about 1 m in this sediment the clay fraction deposited with the sand and silt has a high **electrical conductivity** that conducts the electrical component of the **electromagnetic energy** into the ground, leaving no energy to be reflected back to the surface. No matter what frequency of radar energy was transmitted into this sediment, the waves would be attenuated at that depth. While the GPR literature often states that lower frequency antennae are capable transmitting energy deeper in the ground than those of higher frequency (Annan 2008), when an electrically conductive medium is encountered, radar energy will always be attenuated irrespective of its wavelength.

Weather and Moisture Differences as They Affect Resolution

While scientists pride themselves in having their results repeatable, with GPR there can be daily or even hourly changes that can occur in ground conditions, which can vary datasets in ways that can often make them appear completely different. This need not be a problem if the conditions of the ground that produce radar energy reflections, and therefore produce attenuation or spreading, are understood. In most cases this is not a problem, as diurnal changes or even those from season to season do not often affect the basic chemistry or water saturation in units in the ground. However, when differences are apparent between data collected at different times, explanations are in order.

Water retention and its distribution between buried materials is the most important variable in the production of radar reflections (Conyers 2012, p. 34). Even minor amounts of water added to the ground can sometimes make the difference between being able to see features of interest, or having buried materials remain essentially invisible with GPR. At a historic site in southern Arizona the target of the GPR study was a small irrigation canal used during the 19th century and abandoned sometime in the 1920s. One excavation trench was placed in an area where historic photos showed a low linear surface feature where the vegetation was more pronounced. The hypothesized canal was then excavated and exposed. The base of this small feature is about 60 cm below the ground surface, filled with alternating silt and silty clay laminae. The edges of the canal had alternating thin layers of silt and clay that had been cleaned out of the canals during normal operations and then used to reinforce the margins and banks. The base of the canal had an impermeable clay layer, which is probably the first sediment layer deposited after the canal no longer carried water for irrigation.

A 400 MHz reflection profile was collected just adjacent to the excavations (**Figure 2.7**). When the first profile was collected in December 2013, there had been no significant

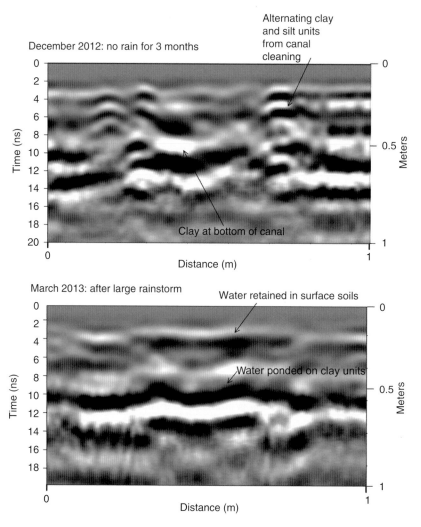

Figure 2.7 Two 400 MHz reflection profiles collected along the same transect 4 months apart. The data collected in December 2012 when the ground was dry shows high amplitude reflections from the canal base and the sediments on its bank, as the clay units retain water and produce velocity discontinuities. After a large rainstorm in March 2013, the water was preferentially retained in surface soils and along a clay unit at about 60 cm and only those units produced visible reflections, making the canal almost invisible.

rainfall for about 3 months and the ground surface was dry. The reflection profile displays reflections from the canal margin units where the sediment clean-out layers are visible as stacked small high-amplitude reflections. The bottom of the canal, where the clay unit had retained water, also produced a high-amplitude reflection. The silty canal fill sediment is mostly non-reflective. After this long period of no rainfall, only the finer-grained clay units retained moisture, with the high permeability coarser grained sediments drained of water. This produced a contrast between these layers with the water retained in the clay having a much slower velocity than the silty and sandier sediments. The change in velocity produced the high-amplitude reflections from the canal base and the layered clay and silt clean-out sediments on the edges.

Just a few days after a large rainstorm in March 2013, the same transect was re-collected with the same antenna and collection parameters as in December 2012. A very different reflection profile was produced (**Figure 2.7**) with a high-amplitude reflection at the base of the surface soil and another one at about 60 cm below the ground surface. The canal edges are only faintly visible and would only have been interpreted as a canal if the previously collected profile was available as a comparison. After water moved into the ground the surface soil retained water and a reflection was generated at the contact of the A horizon and the underlying sediment. Very little water had evaporated from the surface soil prior to data collection, and the soil reflection was visible, where it was not when the ground was dry. A second reflection in the March 2013 profile was produced at about 60 cm, which is the level reached of the downward moving moisture in this permeable ground. Only those two reflections are visible in the profile after the rainstorm. Over time as water continued to drain downward and the surface soil became desiccated, the fine-grained clays on the canal edges would still retain moisture and produce the reflections of interest.

The differential retention and distribution of water can make a great deal of difference in amplitude slice-maps that are produced from profiles collected during times of varying ground moisture. At a site in northern New Mexico a known pit house village was mapped in 2013 with GPR and one structure produced a very distinct circular feature in an amplitude map. Reflective features on the floor of the structure and along its margin produced high-amplitude reflections, as did its central hearth (**Figure 2.8**). The surrounding sediment matrix of this site is a volcaniclastic series of lahar and ash flows. The same structure was tested again in 2014 after a very rainy summer. The same GPR system was used to collect reflection data in the same grid as the previous summer, using the same collection parameters (**Figure 2.8**). Reflection profiles collected after the rains were used to produce an amplitude map using the same sampling and gridding parameters used in the data collected during the dry period, and the known structure is invisible. Only differentially retained pockets of water produced reflections and any reflections from the archaeological features of interest are completely obscured or invisible. This phenomenon is quite common, but in some cases the addition of moisture produces clearer amplitude maps, as water is retained and distributed along boundaries that are of archaeological or geological interest (Conyers 2012, p. 37).

Very different profiles can also be obtained depending on the season of collection. The two profiles in **Figure 2.9** were collected with 400 MHz antennae over the same transect, one in the summer and one when about 10 inches of snow covered the frozen ground surface. The summer profile shows much better resolution of many small bed interfaces in this complex geological setting in Alaska (Conyers 2012, p. 34; 2013, p. 78). In the summer the radar waves moving from the transmitting antenna are much more focused and the conical beam of waves moving into the ground produces reflected waves from much smaller "footprints" as each sediment layer is encountered (see Conyers (2013, p. 66) for equations and tables on spherical spreading of energy). This occurs

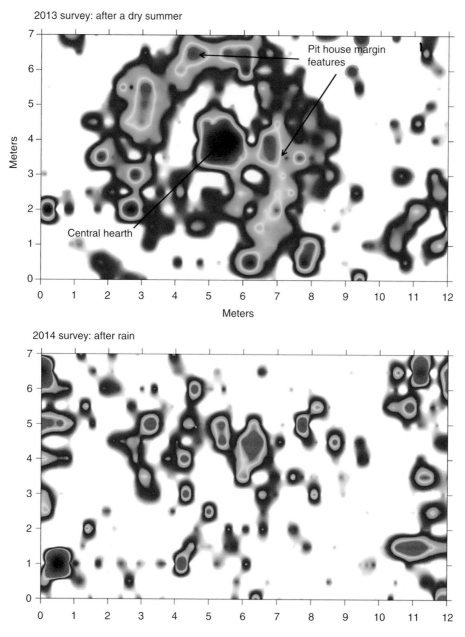

Figure 2.8 Amplitude slice-maps of the same grid collected during two different years using 400 MHz antennae. When the ground was dry in 2013 the outline of the pit house was visible as well as the central hearth. After much rainfall in 2014 only differentially retained pockets of water produce reflections from the same depth in the ground, demonstrating how important it is to take into account moisture in the ground for GPR interpretation. Both maps were from 400 MHz reflections and generated for the 20–50 cm depth.

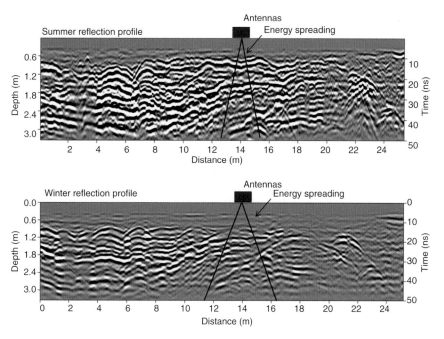

Figure 2.9 Comparison of two 400 MHz reflection profiles collected in Alaska, one in the summer and one in winter when snow covered frozen ground. The energy spreads out more as it travels in frozen ground and produces a more averaged sequence of reflections from the sediment layers. From Conyers (2012). © Left Coast Press, Inc.

because the RDP of moist ground in this area is about 12–13 (Conyers 2013, p. 68) and the radius of the transmission cone at 2 m depth is about 75 cm. With snow cover the transmission cone is much broader as it moves through the snow and underlying frozen ground, which has an average RDP of about 4, which creates a radius of conical transmission of about 1.25 (Conyers 2013, p. 68), spreading the energy out over more area and creating a much more "averaged" reflection profile. As a result of snow cover and frozen ground, there is much less resolution in the stratigraphic units.

The more permeable the ground, the faster changes can occur in reflectivity and depth penetration in the ground as water is added or removed. In coastal Georgia a quartz sand barrier island very near sea level was tested with GPR twice along the same transect, once at low tide and once at high tide. The tidal range can be more than 2 m along this area of Ossabaw Island, and salt water readily moves up and down through the permeable sand in low areas close to the sea level. At high tide the salt water moved upward in the sand so that all radar waves were attenuated within about 40 cm of the ground surface (**Figure 2.10**). The portions of this island that are just 1 m higher in elevation still allowed energy penetration to about 1.2 m depth at high tide, as the salt water had not moved upward to those levels. Along the same transect at low tide, and fortuitously collected after a large rainstorm, the energy penetration along this same transect was about twice as deep, as the salt water table had fallen and fresh water infiltration forced the salty water downward. While the resolution of features in the lower elevation ground is still not good at low tide, even after fresh water infiltration, the reflection profile is still capable of resolving some bedding planes where there was total

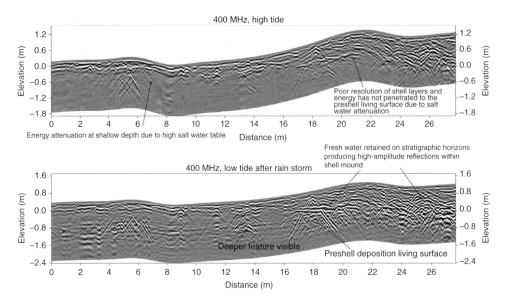

Figure 2.10 Comparison of two 400 MHz reflection profiles collected on a low topography island along the same transect, on the coast of Georgia. The profile collected at high tide shows energy attenuation in the low ground near sea level at a shallow depth due to salt water infiltration and the resulting radar energy attenuation. At low tide, after a rain storm, the salt water was displaced downward and there is deeper energy penetration and greater resulting bed resolution. Data from Victor Thompson, Bryan Tucker, and Matt Golsch.

attenuation at high tide. It is likely that even with fresh water added to the sandy ground, the low topographic areas still retained enough residual salt to produce an electrically conductive medium that affected energy penetration.

References

Annan, Peter (2008) Electromagnetic principles of ground penetrating radar. In: Harry M. Jol (ed.) *Ground Penetrating Radar Theory and Applications*, pp. 3–40. Elsevier Science Limited Kidlington, Oxford.

Baker, Gregory S., Jordan, Thomas E., & Pardy, Jennifer (2007). An introduction to ground-penetrating radar. In: Gregory S. Baker & Harry M. Jol (eds.) *Stratigraphic Analyses Using GPR*, pp. 1–18. Geological Society of America, Boulder, Colorado.

Conyers, Lawrence B. (2012) *Interpreting Ground-penetrating Radar for Archaeology*. Left Coast Press, Walnut Creek, California.

Conyers, Lawrence B. (2013) *Ground-penetrating Radar for Archaeology,* 3rd Edition. Altamira Press, Rowman and Littlefield Publishers, Lantham, Maryland.

Conyers, Lawrence B. & Lucius, Jeffrey (1996) Velocity analysis in archaeological ground-penetrating radar studies. *Archaeological Prospection*, vol. 3, pp. 312–33.

Hollender, Fabrice & Tillard, Sylvie (1998) Modeling ground-penetrating radar wave propagation and reflection with the Jonscler parameters. *Geophysics*, vol. 63, pp. 1933–42.

Jol, Harry M. & Bristow, Charlie S., eds. (2003) GPR in sediments: advice on data collection, basic processing and interpretation, a good practice guide. In: *Ground Penetrating Radar in Sediments*, pp. 9–27. Geological Society Special Publication No. 211, The Geological Society, London.

Neal, Adrian (2004) Ground-penetrating radar and its use in sedimentology: principles, problems and progress. *Earth-Science Reviews*, vol. 66, no. 3, pp. 261–330.

Orlando, Luciana (2007) Georadar data collection, anomaly shape and archaeological interpretation: a case study from central Italy. *Archaeological Prospection*, vol. 14, pp. 213–25.

Van Dam, Remke L., Van Den Berg, Elmer H., Schaap, Marcel G., et al. (2003) Radar reflections from sedimentary structures in the vadose zone. In Harry M. Jol & Charlie S Bristow (eds.) *Ground Penetrating Radar in Sediments,* pp. 257–273. Special Publications, vol. 211, no. 1. The Geological Society, London.

Woodward, John, Ashworth, Philip J. Best, James L., et al. (2003) The use and application of GPR in sandy fluvial environments: methodological considerations. In: Charlie S Bristow & Harry M. Jol (eds.) *Ground Penetrating Radar in Sediments*, pp. 127–42. Geological Society Special Publication No. 211, The Geological Society, London.

Yilmaz, Oz (2001) *Seismic Data Analysis: Processing, Inversion and Interpretation of Seismic Data.* Investigations in Geophysics Number 10, Society of Exploration Geophysicists, Tulsa, Oklahoma.

3 Integration of Geology, Archaeology, and Ground-penetrating Radar

Abstract: When initiating a GPR study in complexly layered ground it is imperative to first correlate reflections visible in profiles to known stratigraphic units in the ground. Velocities of radar waves can be calculated so that waves, recorded in time, can be directly tied to subsurface units that are measured in depth. When geological units are exposed at the surface there can often be a direct correlation of radar wave reflections to layers when antennas are pulled away from the exposures to a study area where they remain buried. These initial steps allow reflections generated from geological layers to be differentiated from those whose origins may be cultural. If it is known what the orientation, depth and composition of both geological and cultural units is in an area, two-dimensional models can be created prior to data collection and then compared to images produced from a survey area. This can be of great value in differentiating geological units from archaeological layers.

Keywords: correlation, sediment packages, isosurfaces, modeling, amplitude slice-maps

Introduction

In all geoarchaeological studies using GPR, once the basics of how radar moves in the ground and reflects off interfaces of interest is understood, a correlation of visible reflections to stratigraphy can proceed. Sometimes referred to as "calibration" of GPR with materials in the ground (Heinz and Aigner 2003), this process can often be confusing and difficult, especially in complexly bedded ground. Another complexity is that GPR reflection data are collected in radar travel time while buried interfaces that might be visible in exposures are measured in distance, making a direct correlation also more difficult. Without open excavations or outcrops that allow a direct correlation of visible stratigraphy to radar reflections in profiles and then additional correlation to many other profiles in a study area, a definition of geological units and their origin and age can also be problematic.

The first step is usually to measure the velocity of radar waves in the ground and then two-way radar travel times of recorded reflections can be converted to depth. This can allow direct correlations between geological units and radar reflections. Details of how to estimate velocity where there are no open excavations for visible correlations, such as **hyperbola fitting** and common midpoint tests, are discussed elsewhere (Conyers

Ground-penetrating Radar for Geoarchaeology, First Edition. Lawrence B. Conyers.
© 2016 John Wiley & Sons, Ltd. Published 2016 by John Wiley & Sons, Ltd.

2013, p. 107). All or some of these velocity determination methods should be employed prior to GPR data collection so that data acquisition parameters can be calibrated for the depth of energy penetration necessary and resolution desired for a particular study (Conyers 2013, p. 88).

Even when depth and velocity calibrations are understood, there is still some question as to which particular horizon in the ground is generating the reflection visible in GPR profiles. As interfaces that produce reflections generate a sine wave that travels back to the surface to be recorded, the first "deflection" in the waveform, as visible in traces, is usually the correct "pick" for the top of a horizon of interest (**Figure 2.2**). Even when this is done, there always remains some question as to whether a planar reflection that is being viewed in GPR profiles was really generated from the top of a particular geological unit of interest. This becomes even more complex if large packages of sedimentary and soil units are being studied, where a direct correlation between many reflections and specific interfaces in the ground is important. If it is only important to identify a series or package of sediment containing many reflections corresponding to multiple stratigraphic units within a depositional package this becomes less important. If so, then the delineation and genesis of each and every reflection is perhaps not a good use of interpretive time as long as general stratigraphic units can be defined.

Examples of Correlating Radar Reflections to Define Stratigraphic Interfaces

There are often situations in the field where visible sedimentary units occur in outcrops adjacent to areas of interest. This type of situation might be where a distinct layer visible in a nearby outcrop is buried by progressively thicker overburden in the area to be surveyed. One simple way to make a direct correlation of radar reflections to interfaces of this sort in the ground is to place the antennae directly on the exposed unit and then collect a reflection profile along a transect that leads from the outcrop into the study area where this horizon is buried. When this is done the unit or multiple units of interest are potentially visible as progressively deeper reflections in the reflection profile (**Figure 3.1**).

An example of this method is from a volcaniclastic sequence in El Salvador, where the unit of interest is a distinctive ancient living surface from about AD 300, which was the horizon on which Mayan people built structures and planted crops (Conyers 1995; Conyers and Spetzler 2002; Sheets 1992). Locally termed the TBJ horizon, it is visible in outcrop as a white, partially cemented ash unit, which was deposited as a flow generated at a volcanic vent many tens of kilometers to the south of the site (Sheets 2002). Not long after the emplacement of this ash, people recolonized this area and built an agricultural village on this surface, which was later buried by many meters of tephra from a nearby volcano (Miller 2002). It is directly on this TBJ living surface that people lived, farmed, and built a village, so its identification in GPR reflection profiles and its three-dimensional mapping in a large unexcavated area were the primary goals of the study (Conyers 1995).

The TBJ ash horizon is visible as a high-amplitude planar reflection in the 270 MHz reflection profile where the antennae were pulled away from its outrop. This ash layer is buried by as much as 3–4 m of tephra in the area of interest (**Figure 3.1**). Not only is the ash easily identified in GPR profiles using this method but stratigraphic work that

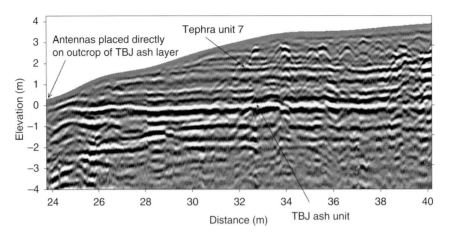

Figure 3.1 A simple way to correlate radar reflections visible in profiles to units of importance is to place the antennae directly on an outcrop and trace them into buried areas of a site. This example is from El Salvador where a volcanic ash unit (TBJ) is buried by many tephra units.

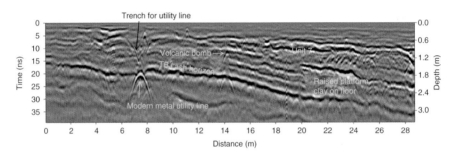

Figure 3.2 Known stratigraphic interfaces that are correlated with GPR reflections can be traced in multiple reflection profiles such as this one from El Salvador. Geological units can be defined and studied, and archaeological features within the stratigraphic package identified.

defines units and their thicknesses in excavations some distance away allows for an identification of other important tephra units. One of these layers is the well-indurated Unit 7, which can be identified in GPR profiles by its high-amplitude reflection.

This integration of GPR reflections with known layers can then be followed up with a more complete analysis of not just these units but also geological and archeological materials within the total volcaniclastic package. In this area of El Salvador other profiles in a grid of reflection profiles were then correlated to this visible stratigraphy and other interesting buried features are visible and can be mapped in three dimensions. In one of these profiles, a distinctive point-source hyperbola from a metal object can be seen directly on the TBJ unit, which is a buried modern utility pipe, with its burial trench visible above it (**Figure 3.2**). Other smaller lower amplitude point-source reflection hyperbolas visible in this profile are likely volcanic bombs that are known to have exploded out of the nearby volcanic vent throughout the eruption. The total eruptive sequence ultimately emplaced 15 separate units of tephra on top of the TBJ (Miller 2002). A clay floor of a Mayan house is visible in the profile (Figure 3.2), which was raised about 70–80 cm above the living surface, consistent with typical Mayan houses excavated nearby (Conyers and Spetzler 2002; Sheets 1992). Cross-beds within an ash flow unit are visible directly on top of the house platform, which sits in the stratigraphic position of Unit 3, a thick ignimbrite unit (Miller 2002). The emplacement of this quickly flowing ash unit undoubtedly carried away the superstructure of this building, a process that has been documented elsewhere at this site (Sheets 1992).

While the archaeological remains encased in these volcaniclastic units were the primary focus of this study, the GPR profiles have enough stratigraphic resolution to also display other interesting geological features (Figure 3.3). In this profile the cross-beds within tephra Unit 3, known to be an ash flow from studies done nearby, are visible. As the profile in Figure 3.3 was collected with 270 MHz antennae, the much thinner Units 1 and 2 that rest directly on the TBJ horizon, which are only about 20–30 cm thick, cannot be resolved. Only the thicker Unit 3 is visible as the propagating wavelength in this material with an RDP of 3 is about 60 cm (Table 2.1). If detailed information about the emplacement history of each of the tephra units was a research goal, a combination of 270 and 400 MHz profiles could be used in conjunction to map the thickness of most of the individual volcaniclastic units in this study area. The 400 MHz energy, however, was only capable of transmitting energy to about 2–3 m in this ground, and therefore the deeper, thinner units within this volcaniclastic package could not be measured and studied.

In other complex geological settings it can be difficult to differentiate the geological strata from archaeological features. This can be particularly difficult in well-layered sediments, such as sand dunes, where each of the aeolian units contains various amounts of clay, some of which retain more water than others, and the interfaces between layers readily reflect radar energy (Figure 3.4). Along the Oregon coast, near the outcrop shown in Figure 1.2, GPR was used to prospect for additional clay floors of pit structures that were thought to exist in the dune sediments. Many interesting reflection profiles were collected and interpreted, all of which displayed the complex units typical of interbedded sand and clayey sand units in this aeolian sequence (Figure 3.4). After topographic corrections, the reflection profiles were used to explore visually for "flat floors" that might be anthropogenic in origin (Figure 3.4). Those planar reflections potentially produced from archaeological features could be identified as they cross-cut dune strata, and therefore were more likely to be anthropogenic features rather than geological in origin. The sediments of interest are dunes that sit on a Pleistocene soil, and all these units were readily visible in the reflection profiles (Figure 3.4).

The identification of a roughly horizontal feature in the upper portion of one sand dune was an important clue to the location of a house floor (Figure 3.4). The floor

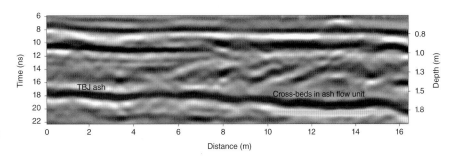

Figure 3.3 Reflection profile from El Salvador displaying the TBJ ash at the base of the stratigraphic sequence of interest, with a cross bedded ash flow above it.

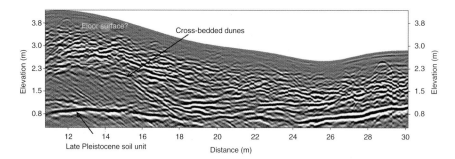

Figure 3.4 Topographically adjusted GPR reflection profiles can display a dizzying amount of reflections, such as this one, using a 400 MHz antenna from coastal Oregon in cross-bedded aeolian dunes. Within this sediment package a horizontal floor surface was visible, which then was analyzed three dimensionally in a grid of profiles.

Figure 3.5 Reflection profile using the 400 MHz antennae showing a horizontal clay floor in sand dune sediment, with many shallow rodent burrows and tree roots producing shallow hyperbolic point-source reflections.

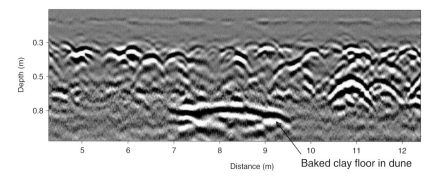

reflection is to some extent distorted vertically in the reflection profile, probably because of minor water retention differences in the overburden sand, which changed the velocity of radar waves enough to distort the horizon (Conyers 2013, p. 158). A grid of GPR data was then collected over this hypothesized floor feature with 400 MHz antennas with 25 cm profile spacing. These reflection profiles readily identify the clay floor and other associated features in these dune sands, with a number of overlying reflection hyperbolas visible, which were likely to have been generated from rodent burrows known to exist in the upper portion of the dunes in this area (Figure 3.5).

When planar reflections of interest are visible in profiles, other visualization methods can be used to produce images of features in three dimensions. One of these tools is called **isosurface** rendering (Conyers 2013, p. 171), which can produce computer-generated visualizations of reflection features with only certain amplitudes of reflections visible, making all other reflections transparent. When this was done with the 400 MHz profiles in a small grid, only the highest amplitude reflections are displayed with two raised platforms on either side of the pit structure floor visible (Figure 3.6).

This method of first identifying reflections visually in reflection profiles followed by the collection of many profiles in a detailed grid directly over the feature for three-dimensional imaging is a logical way to proceed in the discrimination of geological and archaeological reflections. In this way images of only certain amplitudes can be visualized (Figure 3.6). This is a powerful way to identify and study archaeological features encased in complex sedimentary packages (Video 1).

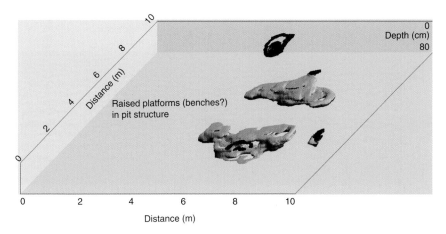

Figure 3.6 Many tightly spaced reflection profiles in a grid can be resampled to produce three-dimensional renderings of only selected amplitudes to define archaeological features of interest in the ground. In this case the raised platforms above the floor of a pit house buried in the aeolian dunes (shown in one profile in **Figure 3.5**) were imaged from coastal Oregon. Moving image of these features is shown in Video 1.

While the identification of anthropogenic features within complexly buried geological settings can be more of an art than science at times, there are other tools that can help identify and interpret archaeological materials in complex geological settings. An example of one of these tools, which was a key to understanding buried prehistoric irrigation canals in fluvial sediment, is two-dimensional reflection profile modeling (Conyers 2013, p. 159). In southern Arizona an area adjacent to a Classic Period Hohokam village was hypothesized to have been the location of prehistoric irrigated maize fields. Today this low-lying, extensive flat surface is preserved on the first fluvial terrace above the modern floodplain, which has been developed into a golf course. This well-tended ground provided a perfect surface on which to collect GPR profiles, but excavations were prohibited, so other means were used to interpret the hypothesized archaeological features visible in reflection profiles. These included first studying these features in two-dimensional profiles to understand the associations of sedimentary units, soils horizons and the canal reflections in space. A synthetic model was then constructed to generate an idealized reflection profile of what these hypothesized archaeological features would look like in actual profiles for comparison to the actual reflection profiles.

The braided stream channels known to exist in this fill terrace sequence could be readily imaged using 400 MHz reflection profiles (**Figure 3.7**). These fluvial channels, and an associated buried soil unit that formed during a period of landscape stability, produce distinct high-amplitude reflections, which can be defined in profile. Many cut and fill surfaces can be seen in the deeper portion of the stratigraphic sequence, as sand and gravel bars were deposited as the braided river was aggrading (**Figure 3.7**). One distinct planar reflection visible in this profile is extensive horizontally, unlike the braided bar units. Nearby outcrops suggest that this unit is a buried soil horizon dating to about 800 years ago. Directly adjacent to this planar surface a distinct vertical "**bow tie**" reflection is visible, with geometry unlike the other typical reflections from fluvial units in this stratigraphic package. Most important is the stratigraphic connection of the top of this unusual "bow-tie" reflection feature to the laterally extensive horizontal reflection, hypothesized to be the buried soil where crops were grown. The connection

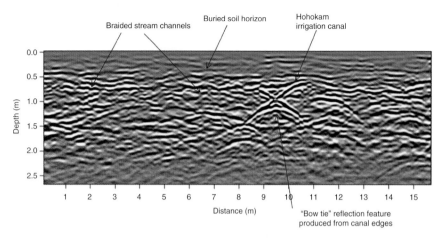

Figure 3.7 A very complex reflections profile, in this case a 400 MHz profile, collected on a braided river terrace from southern Arizona. The prehistoric canal produces a vertical "bow tie" reflection in the profile, which was modeled in **Figure** 3.8 in order to understand its origin. Braided bars produce much less continuous reflections from many cut and fill bed boundaries.

of the "bow tie" and the horizontally continuous planar reflections were hypothesized to be an irrigation canal adjacent to a maize field soil horizon. These were hypothesized to have been constructed, modified, and utilized by the Hohokam people about the year AD 1200.

While the identification of the soil and irrigation channel within the braided stream sediments was considered likely, further documentation of what might have caused the unusual vertical "bow tie"-shaped reflection feature were necessary before an interpretation could be more definitive. One way to support this interpretation was to construct two-dimensional models of GPR reflections using a software modeling program to predict radar wave paths on a computer to generate a model of how buried features would appear in GPR reflection profiles (Conyers 2013, p. 152; Goodman 1994).

In that model a horizontal surface with an adjacent U-shaped canal was simulated to determine how radar waves would travel to and from the surface and intersect the canal and an adjacent soil unit. The computer program then simulated the waves that would return to the surface antenna to be recorded, and their relative amplitude was modeled (Video 2). In this simulation the complexity of an otherwise simple U-shaped canal displays a very different reflection profile (**Figure 3.8**). The base and edges of the canal are recorded in their correct location by radar waves that moved from the surface antenna directly down to be reflected back to the antenna. But as radar energy spreads out in a conical shape from the antenna, propagating waves also encounter the opposite side of the canal by traveling along an oblique pathway. These are reflected from the canal edge and travel back to the surface antenna to be recorded before the antenna has moved directly on top of the canal reflection surface. As the elapsed time is longer for the waves that move in these oblique travel paths, and as the radar system records them as if they were received from directly below, one part of the lower arm of the "bow-tie" is generated. The same phenomenon occurs as the antennae move away from the canal and energy is reflected from the other side of the canal from behind the moving antenna. The result is a vertical "bow tie"-shaped reflection with the upper portion of the feature displaying the correct location of the canal and the lower "artificial" reflections recorded as a function of

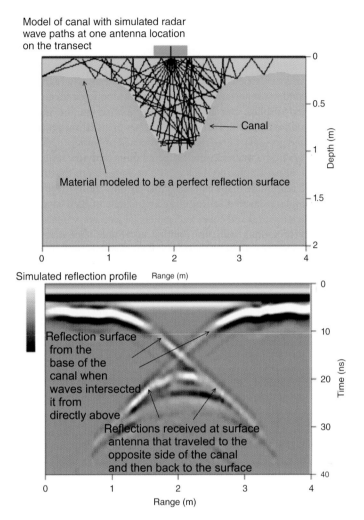

Model of canal with simulated radar wave paths at one antenna location on the transect

Canal

Material modeled to be a perfect reflection surface

Depth (m)

Range (m)

Simulated reflection profile   Range (m)

Reflection surface from the base of the canal when waves intersected it from directly above

Reflections received at surface antenna that traveled to the opposite side of the canal and then back to the surface

Time (ns)

Range (m)

Figure 3.8 Two-dimensional models can be produced on the computer to determine what buried archaeological features would look like under certain geological conditions. In this model a "U-shaped" irrigation canal was modeled, displaying the classic vertical "bow tie" reflection almost exactly similar to that visible in the GPR reflection profile displayed in

Figure 3.7. This reflection feature was created as a function of the spreading of radar waves from the surface antenna that intersect the opposite edges of the canal, with waves moving at an oblique angle, but being recorded as if they were directly below the antenna. Model courtesy of Dean Goodman.

the spreading of energy in the ground and longer travel pathways of energy to and from the opposite canal edges (**Figure 3.8**).

This simulation model of the canal almost perfectly represents the reflection feature visible in the profile adjacent to the hypothesized buried soil horizon (**Figure 3.7**), overlying the braided stream terrace surface. It lends support to the interpretation of its origin. At the very least, this model indicates that the reflection surface visible in the GPR profile is almost the exact shape as that simulated and while this canal-shaped feature has not been excavated to confirm its origin, the model is a very good confirmation.

In addition, in the GPR profile the depth of the top of the canal is exactly the same as a horizontal reflection surface interpreted as the buried soil, which is definitely not part of the braided stream sediments. Both the interpreted soil and associated canal are at exactly the same elevation as a buried soil visible in outcrops nearby.

In many strictly archaeological applications, where the geological matrix of a site is of secondary importance, a large variety of amplitude slice-maps, isosurface renderings, and video displays can be constructed to help in the interpretation of buried features (Goodman and Piro 2013). These types of visualizations are especially helpful if the associated geological materials are composed of homogeneous sediments or the stratigraphic layers in the matrix contain layered units that now reflect high-amplitude waves. Those low or no amplitude areas can be excluded from the archaeological feature reflections, and amplitude maps will likely display only the anthropogenic units of interest. In more complexly layered settings, the abundance of high-amplitude stratigraphic boundaries that surround and cover archaeological features, when sliced and displayed, can confuse interpretations and the resulting amplitude slice-map images can be extraordinarily complex (Conyers 2012, p. 51). If the geological matrix is understood and taken into account, amplitude slice-maps and other images that display the relative strength of reflected waves can still be useful as long as care is taken to differentiate which surfaces in the ground have produced which reflections. Reflections produced from the geological layering must be understood first by interpreting reflection profiles, and then the archaeological reflections, their dimensions, and geometry can be imaged in amplitude maps within the context of the geological matrix.

An example of the necessity to differentiate geological units from archaeological in amplitude slice-maps comes from coastal California where a small stream channel is bounded by cultural features along its bank (Conyers 2012, p. 60). Reflection profiles collected perpendicularly to the channels readily produce two-dimensional images of a fluvial channel in cross section (**Figure 3.9**). The 500 MHz antennae used in this survey produced energy that was attenuated prior to reaching the bottom of the channel. Sediments making up the banks of this small channel are composed of alternating beds of clay and sand, producing interfaces that are highly reflective and distinctly visible in reflection profiles. The stream channel was ultimately filled with homogeneous coarse sand, which is mostly non-reflective.

On the banks of this small creek a number of constructed floors and other built features can be seen in many of the reflection profiles, as well as scattered rocks that were likely brought into this area for building purposes. The banks of this channel were modified for human purposes before the site was abandoned, and later it was filled with sand. The individual rocks produce distinct hyperbolic reflections while the constructed floors generate high amplitude planar reflections (**Figure 3.9**).

To generate amplitude slice-maps the 56 individual reflection profiles in the grid were re-sampled in 5 ns slices, to produce horizontal amplitude maps of about 20 cm thick (**Figure 3.10**). The 5–10 ns slice (from 20 to 40 cm in the ground) shows the edge of the creek at its widest extent as linear high-amplitude features. In progressively deeper slices, the banks of the channel narrow with depth, with the channel fill is represented as low or no amplitude areas between the banks. In the deepest slice displayed, from 15 to 20 ns the constructed features on the east bank of the channel are visible, but would have been difficult to identify only from the amplitude maps. In order to feel confident about the origins of the high-amplitude units visible in these slice-maps, the reflection profiles and the features visible in two dimensions must be interpreted in unison with the horizontal amplitude maps. Only in this way can the geological and archaeological reflections in both two- and three-dimensional images be differentiated.

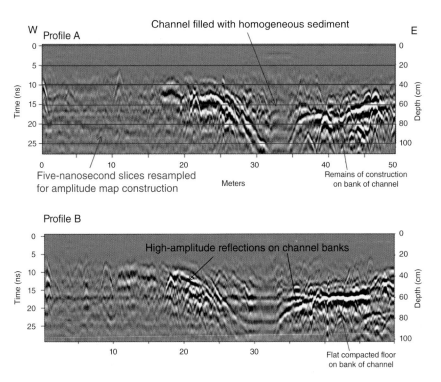

Figure 3.9 Reflection profiles from within a large grid of data in central California collected using 500 MHz antennae. The channel banks, containing archaeological horizons, can be differentiated from geological strata in the profiles as flat high-amplitude planar surfaces. These profiles were sliced digitally in horizontal packages 20 cm (5 ns) thick to produce the amplitude slice-maps in **Figure 3.10**. From Conyers (2012). © Left Coast Press, Inc.

In this chapter a few of the techniques are shown, which can be applied to geoarchaeological studies as a necessary first step to differentiate geological reflections from those constructed or modified by humans. Often, this process in any geoarchaeological GPR study is the most difficult, and can almost never be done as profiles are being collected in "real time," as reflections that are viewed on the GPR system monitors are usually distorted, unprocessed, and difficult to interpret. Reflection profiles must later be computer processed to remove background noise and then be viewed using a variety of display techniques. Often they must also be adjusted for topography in sloping or complex terrains, and the two-dimensional images interpreted first before other displays of reflections in three-dimensions using amplitude slice-maps or isosurface renderings can be understood. Every geoarchaeological study presents new and different conditions, so GPR practitioners must be prepared to conduct a variety of collection and interpretation methods before an understanding of which reflection horizons are of interest, and which are less so. Identifying buried soil units or other geological strata that have archaeological meaning are a crucial first part of this interpretive process. Interpretations about the origin of radar reflections can be aided using modeling programs, but ultimately an identification of buried surfaces in the profiles is necessary before any further work integrating archaeology and geology can commence.

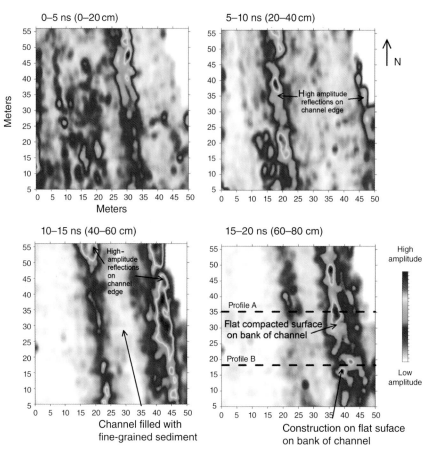

Figure 3.10 Displays of amplitudes of reflected waves in 20 cm thick slices generated from 56 reflection profiles collected in a grid. These maps from central California show the banks of a fluvial channel with archaeological floor features on the east edge. From Conyers (2012). © Left Coast Press, Inc.

References

Conyers, Lawrence B. (1995) The use of ground-penetrating radar to map the buried structures and landscape of the Ceren site, El Salvador. *Geoarchaeology*, vol. 10, no. 4, pp. 275–99.

Conyers, Lawrence B. (2012) *Interpreting Ground-penetrating Radar for Archaeology.* Left Coast Press, Walnut Creek, California.

Conyers, Lawrence B. (2013) *Ground-penetrating Radar for Archaeology*, 3rd Edition. Altamira Press, Rowman and Littlefield Publishers, Lantham, Maryland.

Conyers, Lawrence B. & Spetzler, Hartmut (2002) Geophysical exploration at Ceren. In: Payson Sheets (ed.) *Before the Volcano Erupted: The Ancient Cerén Village in Central America*, pp. 24–32. University of Texas Press, Austin, Texas.

Goodman, Dean (1994) Ground-penetrating radar simulation in engineering and archaeology. *Geophysics*, vol. 59, pp. 224–32.

Goodman, Dean & Piro, Salvatore (2013) *GPR Remote Sensing in Archaeology*. Geotechnologies and the Environment, Volume 9, Springer Science, New York.

Heinz, Jürgen & Aigner, Thomas (2003) Three-dimensional GPR analysis of various Quaternary gravel-bed braided river deposits (southwestern Germany). In: Charlie S. Bristow & Harry M. Jol (eds.) *Ground Penetrating Radar in Sediments*, pp. 99–110. Geological Society Special Publication No. 211, The Geological Society, London.

Miller, C. Dan. (2002) Volcanology, stratigraphy, and effects on structures. In: Payson D. Sheets (ed.) *Before the Volcano Erupted: The Ancient Cerén Village in Central America*, pp. 11–23. University of Texas Press, Austin, Texas.

Sheets, Payson D. (1992) *The Ceren Site: A Prehistoric Village Buried by Volcanic Ash in Central America*. Harcourt Brace Jovanovich, Fort Worth, Texas.

Sheets, Payson D., ed. (2002) *Before the Volcano Erupted: The Ancient Cerén Village in Central America*, University of Texas Press, Austin, Texas.

4

Fluvial, Alluvial Fan, and Floodplain Environments

Abstract: Rivers, river terraces, and alluvial fans were locations commonly associated with human occupations and other activities, and are therefore environments of sedimentary deposition that are often analyzed geoarchaeologically. Fluvial sequences in braided and meandering rivers and streams can produce complex GPR reflection profiles exhibiting both erosion surfaces and depositional bedding features. Placing archaeological materials within these environments with GPR analysis can show how people adapted to these changing river systems. River terraces that contain the sedimentary record of the past river systems and are now raised above active floodplains were common habitation areas. Soils and sediments on and below these surfaces contain important buried sites. The erosion and burial agents common along alluvial fans and other high energy depositional environments are also visible in profiles and these sediments can be discriminated from archaeological materials.

Keywords: alluvial, fluvial, debris flows, river terraces, floodplain, canals, channel variations, point-bar, braided channels, landscape analysis

Introduction

Rivers and associated floodplain depositional environments are important environments associated with archaeological materials as people have been drawn to rivers throughout time. These environments visible with GPR are channel systems composed of sand and gravel in complex channel belts within braided rivers, finer grained sediments within meandering fluvial systems, and a wide range of associated environments on floodplains (Tolksdorf et al. 2013). Placing humans within these complex depositional settings can be difficult, as many artifacts can be out of place (Butzer 1982), having been eroded and redeposited as clasts along with the sediments. Meandering river systems where sediments were deposited in lower gradient areas, and therefore composed of finer grained more cohesive sediments that are not as easily eroded, can potentially contain archaeological materials that have a greater likelihood of being in place (Waters 1992). Aggrading floodplains adjacent to rivers can be good depositional settings for the preservation of archaeological materials where fine-grained low-velocity sediments are interbedded with soil units that formed between floods during periods of relative landscape stability. Within actively aggrading river systems and associated flood-prone

Ground-penetrating Radar for Geoarchaeology, First Edition. Lawrence B. Conyers.
© 2016 John Wiley & Sons, Ltd. Published 2016 by John Wiley & Sons, Ltd.

areas, archaeological sites associated with farming, hunting and gathering, and other activities of a less permanent nature are common (Brown 1997).

An analysis of fluvial systems using GPR within an aerially extensive landscape analysis can potentially define and map individual channel bars, identify portions of a fluvial sediment package that were deposited during changing flow regimes, and their analysis can potentially show how water flow in a river might have changed seasonally or over decades or centuries (Brown 1997). The scale of river channels, their depths and migration across a floodplain over time, and the locations of those changing environments can then be mapped spatially, and human use and possible modifications of those environments studied. An analysis of this type, incorporating GPR with more standard geoarchaeological research, could be used to analyze flood frequencies in the past and potentially changes in landscapes along the length of a river system (Brown 1997; Nials et al. 2011).

Other fluvial environments included within the general topic of river systems are river margin terraces that were common locations for human habitation as they were near floodplain and other river resources, but above the level of all but the most unusual floods (Waters 1992). Burial mechanisms on raised river terrace surfaces are usually less dynamic than in the active floodplain with preservation of archaeological materials mostly limited to soil buildup, anthropogenic burial, or rare sedimentation episodes from unusually dynamic floods (Waters 1992). Artifact densities and the types of sites that might be found in all of these fluvial environments can potentially be predicted if the flow regimes can be understood, with high water flow rates leading to low artifact density in mostly out of place conditions (Brown 1997).

Alluvial fans also contain fluvial channels, which are often complexly braided, with steep-sided channels in their upper reaches and smaller scale braided channels near their toes (Nials et al. 2011). These alluvial systems were common environments for humans who lived on the fan surfaces, sometimes developed the channels for farming, and exploited other resources nearby.

All fluvial systems, other than the abandoned floodplain and associated environments on raised terraces, are dynamic environments where the locations of human activities are a function of landscape changes. River locations can vary often with lateral and vertical erosion (degradation), deposition (aggradation), and longer periods of stability where soil horizons developed in some locations. All these geological variations within ancient landscapes (Heinz and Aigner 2003) can be analyzed with GPR in three dimensions, which can place sediment and soil horizons within the context of human exploitation and modification.

Fluvial Systems

A differentiation of fluvial bar types and their geometry in braided and meandering systems can be difficult using GPR in the same way that complexity is noticeable when viewing exposures of these sediment packages in outcrops or excavations (Mumphy et al. 2007). All fluvial sediments contain large scale sedimentary structures that are commonly visible with GPR including cross bedding, cut and fill structures, and erosional contacts. The complexity of fluvial units with a variety of reflections from many types of sedimentary contacts have led some GPR interpreters to attempt to define GPR "reflection patterns" in profiles that could presumably define fluvial facies (Mumphy et al. 2007). These patterns are a series of reflections from bedding planes and sediment changes, described in terms such as parallel, sub-parallel, hummocky, contorted, and cut-and-fill. While these elements that describe the way GPR profiles define bedding contacts can be important when interpreting fluvial systems, by themselves they do not necessarily define ancient environments (Hickin et al. 2007). They can be useful, however, as they are an indication of the river flow regimes that produced individual units visible with GPR, which are helpful in defining environments of deposition. Some of those useful types of descriptions with respect to GPR reflections will be used to describe the examples below.

In the Brahmaputra River floodplain of Bangladesh (Best et al. 2003; Bristow 1993) many types of fluvial sedimentary packages are visible in a reflection profile with very good depth penetration and resolution (**Figure 4.1**). Two packages of sediments are visible in this reflection profile with a lower unit that is about 2 m thick consisting of a cross-bedded, primarily sand unit resting directly on a basal truncation surface. This lower bar contains primarily well-winnowed sand with some weak cross-beds visible, but few laminae that would have produced high-amplitude radar reflections. These types of units are typical of point-bar deposition that accrete laterally during bar migration, especially during high water flow in meandering channels.

Along this area of the Brahmaputra River, water flow changes dramatically seasonally with monsoon flooding followed by lower flow during the dry season. In monsoon floods rapid channel migration deposits sediment in laterally migrating point-bar units, and when water flow decreases a different facies composed of compound bars in braided channels are deposited (Best et al. 2003). This reflection profile displays both these types of fluvial units. The uppermost sequence composed of small cut and fill sequences is where smaller scale channel units were deposited as low-flow braided bars. These were interbedded with finer-grain sediments that produce distinct reflections, with each interface in these small bars reflecting radar energy (**Figure 4.1**).

It is unlikely that the high and low energy bars visible in the fluvial sequence in **Figure 4.1** contain any archaeological materials that are in primary context. All the features visible in this profile are the product of the movement and redeposition of clastic materials by flowing water with different flow rates. The GPR reflection profiles in this environment are useful for understanding the flow in this river system and possibly its changes over time, which would be important in placing these fluvial sequences into a larger landscape analysis.

Smaller scale channel features where erosion and periodic deposition occur infrequently and with a minimum velocity are features within ancient environments that have always drawn people. In small-scale channels water is available for human purposes

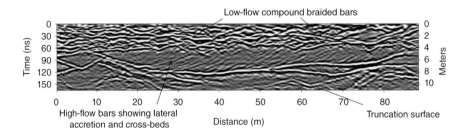

Figure 4.1 Reflection profile using 100 MHz antennae in the Brahmaputra River of Bangladesh displaying two types of river units. The high-flow point bars produced by rapid lateral accretion are well-sorted sand units that are cross-bedded, resting on a basal truncation surface. The lower flow braided bars display many cut and fill features common to complexly braided fluvial systems with alternating poorly sorted sediments. Data from Charles Bristow.

as well as a large variety of plant and animal resources. One of these very small-scale channels was visible in an outcrop in coastal Portugal where late Pleistocene age fluvial and aeolian sediments filled channels carved into Jurassic age bedrock (**Figure 4.2**). At the time people were exploiting this environment the bedrock surface had been scoured and gullied as many small streams flowed from an eroding highland area to the west. Very close to the Jurassic bedrock surface, as these channels were starting to fill with sediment, people were hunting and sharpening stone tools in what has been described as a short-term hunting camp during the late Paleolithic period (Bicho and Haws 2012; Conyers et al. 2013). A GPR survey was conducted to place these artifacts within a wider fluvial landscape analysis.

In order to first understand how the Jurassic surface appears in GPR profiles a direct correlation was made with a GPR transect collected directly on top of an outcrop where the channels are partially exposed (**Figure 4.2**). In a 400 MHz reflection profile these small channels are very distinctive with a high-amplitude reflection generated from the top of the Jurassic bedrock. The sand that fills this channel is reworked aeolian sediment that was likely derived from coastal dunes just to the west, where the headwaters of this small drainage are located. This fine-grained sand displays no reflections whatever, as it is homogeneous and there are no interfaces such as cross bedding to reflect energy (**Figure 4.3**).

The amplitude of the top Jurassic surface is variable along the length of the reflection profile due to irregularities in that surface (**Figure 4.3**). Point-source reflections were generated from small promontories on the bedrock surface, similar to how individual stones would appear in a GPR reflection profile. Where the Jurassic surface is flat and

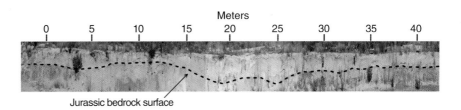

Figure 4.2 Outcrop of a thin fluvial sand sequence resting on an unconformity with Jurassic bedrock, used as a test of GPR profiles shown in **Figure 4.3**.

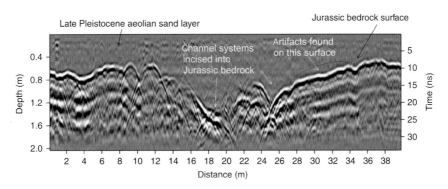

Figure 4.3 A 400 MHz reflection profile collected adjacent to the outcrop shown in **Figure 4.2**, showing the small fluvial channels incised into the Jurassic bedrock surface. This profile was collected in order to identify the bank of a channel where late Paleolithic artifacts were discovered nearby.

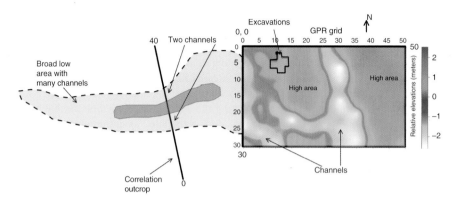

Figure 4.4 Map view of the late Pleistocene fluvial channel system incised into the Jurassic bedrock shown in **Figures 4.2** and **4.3**. The artifacts in the excavations were placed into this paleoenvironmental context as a way to interpret human behavior in this ancient landscape.

regular the interface is displayed as a high-amplitude planar reflection. Nearby the Paleolithic artifacts were found in the sandy sediment, just above a flat surface of this sort, a logical place for a camp and work site. Distinct bedrock knobs sticking up into channel fill dispersed radar energy that encountered the interface, sending radar waves away from the surface antennae and were not recorded. Radar waves reflected from planar bedrock surfaces that face away from surface antennae exhibit a low-amplitude planar reflection (Conyers 2013, p. 52).

A grid of 400 MHz reflection profiles was then collected about 20 m east of the correlation outcrop (**Figure 4.4**) where the channels could be mapped in proximity to the excavations where the stone tools were discovered (Conyers et al. 2013). The elevation of the base of this small channel was then measured directly from the GPR reflection profiles and the surface mapped throughout the grid. The channel system visible in the outcrop shows two small channels that merge with a more complicated anastomosing channel system to the east (**Figure 4.4**). Two distinct topographic high areas between the channels are visible in the buried topography map, rising about 2–3 m above the bottom of the channels. On the northwest flank of one of these small raised areas, on a flat surface, the camp and work site was discovered.

The more extensive floodplain that this small channel system flowed into was located about a kilometer to the east of this site (Conyers et al. 2013). These small topographic rises mapped with GPR were where bedrock features likely provided shelter from the wind and a location where hunters could remain obscured from game in the floodplain to the east. The long reflection profile, which shows a lake and sand dunes in this floodplain discussed above (**Figure 1.2**), is located about 500 m east of this location. While these two GPR studies (**Figures 1.2** and **4.4**) have not been integrated into a larger landscape analysis, they show how complex this late Pleistocene environment was with many fluvial units, lakes, and sand dunes. A small coastal range covered in sand dunes was located to the west, and small channels drained those high areas, flowing eastward into the floodplain. The ground surface was a fairly rugged area with numerous erosional channels bounded by topographic highs of exposed bedrock.

The GPR mapping shows that this location was specifically chosen by Paleolithic hunters as a protected camping place where many tasks were performed as they prepared for hunting game in the floodplain just to the east. Lakes and dunes were located on the nearby floodplain (Figure 1.2), around which there were a variety of animal and plant resources that drew people to this area.

The GPR mapping of this small channel system, as well as the information from the longer GPR profiles to the east, provides a way to place this otherwise marginally interesting lithic scatter within a landscape context to provide much more meaning when interpreting ancient human behavior (Conyers et al. 2013). Only a direct correlation of the small channels visible in one outcrop and then projecting those fluvial features into the subsurface with GPR, allows for a greater understanding of this late Pleistocene landscape.

Fluvial Terraces

When fluvial environments are preserved in river terraces above the active floodplain (Waters 1992) they can be the locations of human habitation sites and other activity areas. On these geomorphic features the relatively flat treads are usually stable environments where soils form and there can be aeolian deposition, while the terrace scarps (risers) can still be active areas of deposition and preservation of archaeological materials (Kuehn 1993). Scarps of terraces were the locations of kill sites (Holliday 1987; Rains et al. 1994), irrigation canals (Beckers and Schütt 2013; Nials et al. 2011), and places where trash was discarded.

On the terrace tread of the San Gabriel River in southern California a number of historical sites are preserved dating to the period when this area was part of Mexico (Conyers 2012, p. 62). A number of GPR grids of profiles were collected from a location on the terrace tread, along the eroded scarp and into the floodplain of the river. Today, the floodplain is inactive due to recent damming and channel enclosure. The topographically corrected reflection profiles (Figure 4.5) display a variety of reflections from fluvial units, both in the modern floodplain and on the terrace. The fine-grained alluvium in the modern floodplain is largely unreflective probably due to bioturbation and thick soil formation.

When the reflection profiles were interpreted there was one area along the scarp where an abundance of small reflection hyperbolas were visible, concentrated right along the terrace edge (Figure 4.5, Profile B). The origin of these small point sources was not immediately understood, but when amplitude slices were constructed (Figure 4.6), the concentration could be seen to follow the terrace scarp. Even without mapping the subtle topographic expression of the terrace edge, the modern floodplain could be differentiated visually from the terrace sediments in the amplitude slice-maps by the arcuate linear boundary (Figure 4.6). When the concentration of these hyperbolas was interpreted within the context of the fluvial landforms, it was apparent that this was likely a trash dump (midden) where objects were discarded into the active floodplain. One excavation was placed here and many historic objects were recovered in this anthropogenic deposit (Conyers 2012, p. 62).

Floodplains and associated fluvial terraces can contain many archaeological materials, which can be difficult to differentiate in GPR profiles from the abundance of reflections usually visible from fluvial channel boundaries (Waters 2008). Canals used for irrigation have been studied using GPR (Carrozzo et al. 2003; Hruska and Fuchs 1999; Sandweiss et al. 2010; Sternberg and McGill 1995) with varying success. They can appear much like other fluvial channels when viewed in cross section, and both anthropogenic and natural features of this sort will often produce similar cut and

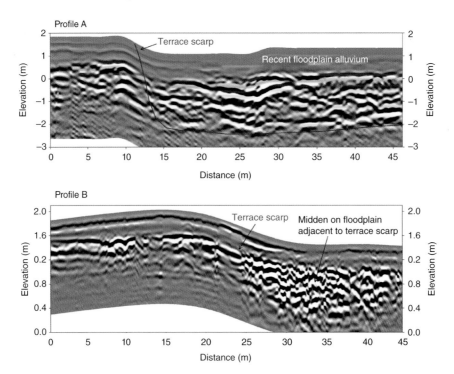

Figure 4.5 Reflection profiles using 400 MHz antennae across a fluvial terrace scarp in southern California scarp displaying floodplain sediments and fluvial units. An artifact midden from historic times is visible as many point-source hyperbolas from objects, deposited along the scarp and buried by floodplain sediments.

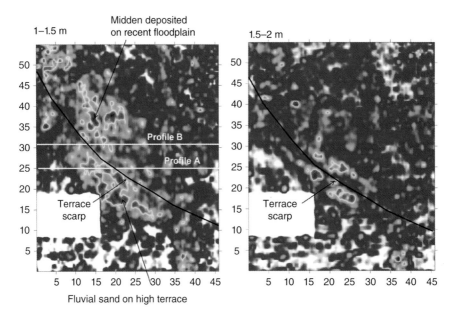

Figure 4.6 Amplitude maps of the fluvial and terrace sediments in southern California showing the terrace edge and the many small high-amplitude reflections from objects in a trash midden deposited on the floodplain along the terrace scarp.

Alternating thin beds of sand and silt
that filled canal after abandonment

Canal edge

Clay in canal bottom

Figure 4.7 Photograph of the complex layering within a canal fill unit from southern Arizona. From Conyers (2012). © Left Coast Press, Inc.

fill sequences and sometimes the vertical "bow tie" effects of complex radar wave travel paths (**Figure 3.8**).

In southern Arizona a fluvial terrace above the Santa Cruz River contains a many irrigation canals and associated agricultural fields that may date to 2500–3000 BP (Huckleberry and Rittenour 2014). These canal systems are very complex, showing evidence of periodic damage from flooding, reconfiguration, and then more use, finally filling naturally after they were abandoned (Huckleberry 1999). In addition, they are preserved within a sequence of fluvial river sediment, and were periodically eroded during dramatic floods, where natural fluvial channels were incised through the irrigation canals. This complexity, and the fact that the GPR data were acquired during active excavations (Huckleberry 2011), provides a good test of the use of GPR to differentiate anthropogenic from natural channels.

The base of the irrigation canals often contains a thin layer of clay, which provides an excellent lithological contrast with the surrounding silt and fine-grained sand units. In addition, the more vertical edges of the canals also provided a reflection surface (**Figure 4.7**). Elsewhere in the vicinity of this study in southern Arizona, the lack of contrast in sediment types between the canal fill and the surrounding material made the canals almost invisible as no high amplitude reflections were generated. This lack of contrast has been described elsewhere, and can be very common not just in irrigation canals but also in fluvial systems where sediment types are unimodal in grain size (Valdés and Kaplan 2000).

The irrigation canals in the study area were filled with thin layers of alternating clayey silt and silty sand, deposited during normal irrigation operations, and especially after abandonment when they filled naturally with sediment. Those thin layers of canal fill were effectively invisible with the 400 MHz energy that was transmitted into the ground, as the units were not thick enough to be resolved.

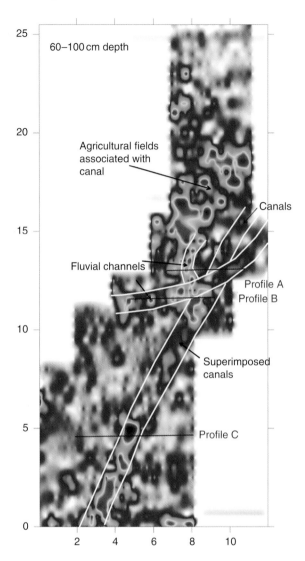

Figure 4.8 Amplitude slice-map from 60 to 100 cm depth showing sediments within irrigation canals with associated fluvial channels that were eroded through the canals during flood periods. Prehistoric agricultural fields were watered by these canals on a river terrace in southern Arizona. From Conyers (2012). © Left Coast Press, Inc.

An amplitude slice through an area where a number of canals, natural fluvial channels, and associated agricultural fields were located shows a wealth of high-amplitude reflections that cannot be used to readily delineate the canals of interest (**Figure 4.8**). Each reflection profile had to be interpreted in two dimensions in order to try to understand what was producing the complicated reflections visible in map view (**Figure 4.9**). The image in Profile C (**Figure 4.9**) across the southern portion of the irrigation canal displays a modified vertical "bow-tie" reflection feature where very high-amplitude reflections from the canal surface on the "opposite side" generate high-amplitude reflections as those surfaces were perpendicular to the incoming radar waves. The energy that moved directly down to the canal and back to the surface was mostly scattered away

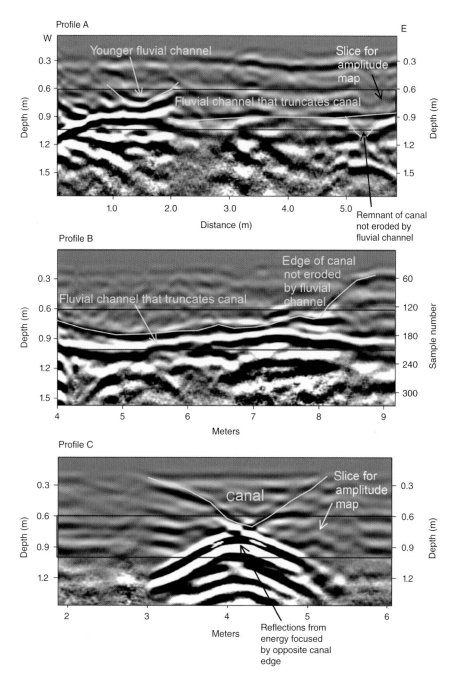

Figure 4.9 Reflection profiles showing the complexity of canal and fluvial channel reflections in southern Arizona using 400 MHz antennae. These profiles are placed within the grid in Figure 4.8. It is difficult to differentiate the canals from fluvial channels in this package of materials unless each profile is interpreted individually and the origins of reflections are analyzed.

from the surface antenna, and so little of it was reflected back to the surface. The resulting image shows the canal edge from directly below the antenna to be barely visible (**Figure 4.9**, Profile C). The amplitude slice that crosses this reflection feature in the 60–100 cm depth (**Figure 4.8**) is displaying the lower portion of the "bow-tie" feature as very high-amplitude reflections.

Farther to the north along this canal there are three separate canals, each of which was rebuilt as part of a previous water distribution system partially destroyed by flooding. These are also truncated by a natural fluvial channel that eroded into the irrigation canals (**Figure 4.9**, Profile B). One reflection profile was collected parallel to this natural channel, and its bottom appears as a distinct planar reflection (Profile B in **Figure 4.9**). These multiple reflections create some of the confusion in the amplitude slice-map, as this fluvial channel, some of the remnants of the irrigation canals, and the variability of sediment types that were preserved in all reflect energy in different ways at different locations (**Figure 4.8**). To the north in Profile A (**Figure 4.9**) this situation becomes even more complex as the fluvial channel appears to cut through the canals. Coarse sediment from a smaller channel was deposited in the agricultural field adjacent to the canal locations, producing a wealth of high-amplitude reflections there (**Figure 4.8**).

This example illustrates the complexity of channels and canals when they are interbedded, truncate each other, and each contains variable amounts and grain sizes of sediment. The complexity of reflections from these features in amplitude maps will often belie interpretation based on the geometry of the amplitude features alone. Even when interpretation of the individual reflection profiles is made, the complexity is almost beyond usefulness. It was only possible here because detailed excavations in trenches paralleling many of the profiles allowed an integration of visible sedimentary layers with the GPR reflections (Huckleberry 2011).

Alluvial Fans

Along the toe of an alluvial fan that was deposited on a raised river terrace of the Santa Cruz River in southern Arizona a Classic Period Hohokam mound is preserved. This feature is unexcavated and appears only as a very low-relief mound with scattered pottery fragments and multiple layers of eroded clay that was used as construction material for what were above-ground room blocks. Once these buildings were abandoned, perhaps sometime around AD 1400, the adobe "melted" into layers around a few still-standing walls, until all the exposed walls were finally eroded and covered. Sometime after abandonment, flood waters along many braided channels of the alluvial fan flowed around the mound, eroding parts of it.

A reflection profile that crosses a part of the adobe mound shows well-layered units of the adobe melt, which is all that remains near the surface of what was once standing architecture. This mound creates a topographic rise, and the flood waters that coursed through this area only eroded its very edges (**Figure 4.10**). The small scale channels on the alluvial fan are visible in all the reflection profiles collected in this grid of 400 MHz data. They are easily interpreted as channels, and can be discriminated from the subhorizontal planar reflections produced from adjacent adobe melt layers.

An amplitude map of the grid over the western portion of the mound shows the linear channels filled with sand wrapping around the mound (**Figure 4.11**). Other amplitude features within the mound itself are not easy to interpret, as they might have been generated from adobe melt layers, or possibly some intact walls or floors of the original building.

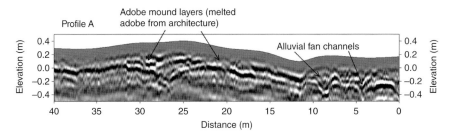

Figure 4.10 A 400 MHz reflection profile showing small alluvial fan channels that incised into the toe of a fan near a site containing adobe melt layers of architecture in southern Arizona.

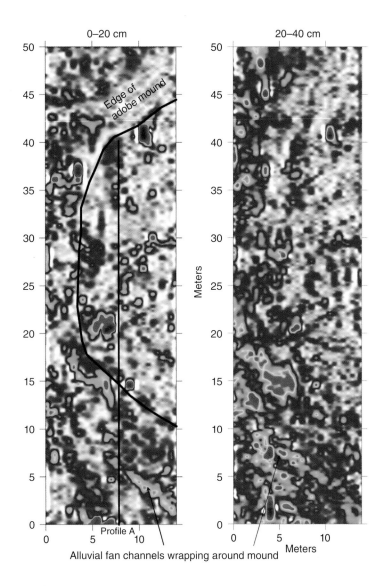

Figure 4.11 Amplitude slice-maps showing the small alluvial fan channels wrapping around adobe mound archaeological site in southern Arizona.

In highland Ecuador an Inca temple, which may be that of the last Inca emperor at the time of Spanish conquest, has been exposed in excavations going back many decades (Bray and Almeida 2014). The beautifully cut stone of the temple and associated elite residences stands in stark contrast with the sediments that buried it (Figure 4.12), which were deposited as alluvial fan channels and debris flows. In the unexcavated area of the site, adjacent to the temple stones, the alluvial stratigraphy composed of interbedded sand and cobble layers appears in GPR reflection profiles as poorly defined planar reflections, where cobble-rich layers sit on finer grained sandy channels (Figure 4.13).

Only the larger cobbles in this otherwise course-grained, poorly sorted deposit of alternating alluvial sediments and debris flow deposits produced reflection hyperbolas (Figure 4.13). In one area where a larger channel is filled with a unit of cobbles, the cobble bed–sand interface does not produce a planar reflection. This is likely because the larger cobbles in the channel have reflected and scattered the 400 MHz radar waves away from the surface, and impede any remaining energy from reaching the lower sand–cobble interface. This confusion of multiple radar reflections in coarse grained, poorly sorted units of this sort is common in high-energy deposits of this sort. Higher frequency antennae could potentially produce reflections from each and every cobble in these sequences for higher definition but the result could be quite "busy."

Using only the 400 MHz reflection profiles collected in this area, an amplitude slice-map through an unexcavated area displays the distinct square edges of a still buried building, which is likely associated with the Inca temple nearby (Figure 4.14). The surrounding debris flow and alluvial matrix that cover it can be easily differentiated in map view as areas displaying only random reflections from the larger clasts in these poorly sorted alluvial sediments.

Once the alluvial materials are correlated to the exposures of these coarse-grained sediments that provide the matrix of the site, reflection profiles can be used to interpret the cultural and geological materials in this setting (Figure 4.15). The distinct

Figure 4.12 Cut-stone architecture in an Inca palace that is covered by alluvial fan and debris flow sediments in northern Ecuador.

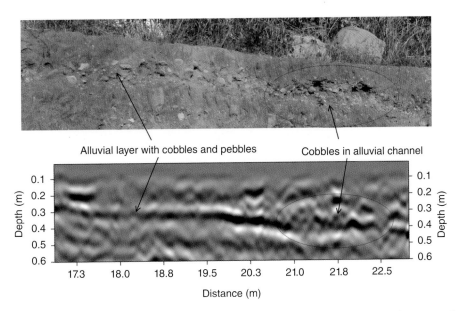

Alluvial layer with cobbles and pebbles   Cobbles in alluvial channel

Figure 4.13 A 400 MHz reflection profile in this alluvial fan deposit is barely able to define the small clasts in the cobble and pebble unit. The interface between the units of different clast sizes could be defined only in locations where the radar energy was not scattered by an overlying thick layer of cobbles.

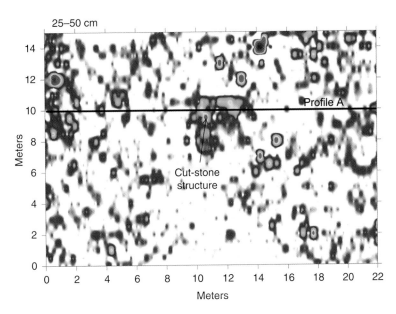

Figure 4.14 Amplitude slice-map generated from 400 MHz reflection profiles showing a buried stone structure similar to the one excavated and shown in **Figure 4.12**. The surrounding alluvial fan and debris flow sediment produced only scattered point-source reflections generated from the larger clasts in the deposits.

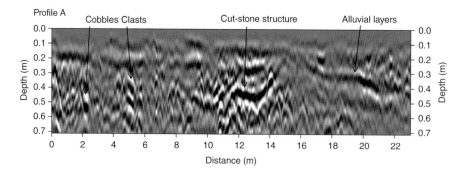

Figure 4.15 A 400 MHz reflection profile showing point-source reflections from clasts and some low-amplitude alluvial fan strata surrounding an Inca stone structure in northern Ecuador.

stone structure displays high-amplitude planar reflections from its interfaces with the surrounding sediment. A few of the alluvial units and many hyperbolic reflections from the larger cobbles or small boulders in this environment are quite distinct.

References

Beckers, Brian & Schütt, Brigitta (2013) The chronology of ancient agricultural terraces in the environs of Petra. In: Michael Mouton & Stephan G. Schmid (eds.) *Men on the Rocks: The Formation of Nabataean Petra*, pp. 313–50. Logos Verlag, Berlin.

Best, James L., Ashworth, Philip J., Bristow, Charles S. et al. (2003) Three-dimensional sedimentary architecture of a large, mid-channel sand braid bar, Jamuna River, Bangladesh. *Journal of Sedimentary Research*, vol. 73, no. 4, pp. 516–30.

Bicho, Nuno & Haws, Jonathan (2012) The Magdalenian in central and southern Portugal: human ecology at the end of the Pleistocene. *Quaternary International*, vol. 272, pp. 6–16.

Bray, Tamara L. & Almeida, José Echeverría (2014) The late imperial site of Inca-Caranqui, northern highland Ecuador: at the end of empire. *Ñawpa Pacha*, vol. 34, no. 2, pp. 177–99.

Bristow, Charles S. (1993) Sedimentary structures exposed in bar tops in the Brahmaputra River, Bangladesh. In: James L. Best & Charles S. Bristow (eds.) *Braided Rivers*, pp. 277–89. Geological Society of London Special Publication 75, Geological Society, London.

Brown, A.G. (1997) *Alluvial Geoarchaeology: Floodplain Archaeology and Environmental Change*. Cambridge University Press, Cambridge.

Butzer, Karl W. (1982) *Archaeology as Human Ecology: Method and Theory for a Contextual Approach*. Cambridge University Press, Cambridge.

Carrozzo, Maria Teresa, Leucci, Giovanni, Negri, Sergio, & Nuzzo, Luigia (2003) GPR survey to understand the stratigraphy of the Roman Ships archaeological site (Pisa, Italy). *Archaeological Prospection*, vol. 10, no. 1, pp. 57–72.

Conyers, Lawrence B. (2012) *Interpreting Ground-penetrating Radar for Archaeology*. Left Coast Press, Walnut Creek, California.

Conyers, Lawrence B. (2013) *Ground-penetrating Radar for Archaeology*, 3rd Edition. Altamira Press, Rowman and Littlefield Publishers, Lantham, Maryland.

Conyers, Lawrence B., Daniels, J. Michael, Haws, Jonathan A., & Benedetti, Michael M. (2013) An upper Palaeolithic landscape analysis of coastal Portugal using ground-penetrating radar. *Archaeological Prospection*, vol. 20, no. 1, pp. 45–51.

Heinz, Jürgen & Aigner, Thomas (2003) Three-dimensional GPR analysis of various quaternary gravel-bed braided river deposits (southwestern Germany). In: Charles S. Bristow & Harry M. Jol (eds.) *Ground Penetrating Radar in Sediments*, pp. 99–110. Geological Society Special Publication No. 211, The Geological Society, London.

Hickin, Adrian S., Bobrowsky, Peter T., Paulen, Roger C., & Best, Mel (2007) Imaging fluvial architecture within a paleovalley fill using ground penetrating radar, Maple Creek, Guyana. In: Gregory S. Baker &

Harry M. Jol (eds). *Stratigraphic Analysis Using GPR*, pp. 133–54. Geological Society of America Special Papers 432, Boulder, Colorado.

Holliday, Vance T. (1987) Geoarchaeology and late quaternary geomorphology of the middle South Platte River, northeastern Colorado. *Geoarchaeology*, vol. 2, no. 4, pp. 317–29.

Hruska, Jiri & Fuchs, Gerald (1999) GPR prospection in ancient Ephesos. *Journal of Applied Geophysics*, vol. 41, no. 2, pp. 293–312.

Huckleberry, Gary (1999) Stratigraphic identification of destructive floods in relict canals: a case study from the middle Gila River. *Kiva*, vol. 6, pp. 7–33.

Huckleberry, Gary (2011) Canals within the Pima County Interconnect Project Corridor, Arizona. In: Douglas B. Craig (ed.) *Archaeological and Geoarchaeological Investigations Along the Santa Cruz River Floodplain: The Pima County Plant Interconnect Project, Technical Report 09–47*, pp. 125–58. Northland Research Inc., Tempe, Arizona.

Huckleberry, Gary & Rittenour, Tammy (2014) Combining radiocarbon and single-grain optically stimulated luminescence methods to accurately date pre-ceramic irrigation canals, Tucson, Arizona. *Journal of Archaeological Science*, vol. 4, pp. 156–70.

Kuehn, David D. (1993) Landforms and archaeological site location in the little Missouri badlands: a new look at some well-established patterns. *Geoarchaeology*, vol. 8, no. 4, pp. 313–32.

Mumphy, Andrew J., Jol, Harry M., Kean, William F., & Isbel, John L. (2007) Architecture and sedimentology of an active braid bar in the Wisconsin River basin on 3-D ground penetrating radar. In: Gregory S. Baker & Harry M. Jol (eds.) *Stratigraphic Analysis Using GPR*, pp. 111–32. Geological Society of America Special Papers 432, Boulder, Colorado.

Nials, Fred L., Gregory, David A., & Hill, J. Brett (2011) The stream reach concept and the macro-scale study of riverine agriculture in arid and semiarid environments. *Geoarchaeology*, vol. 26, no. 5, pp. 724–61.

Rains, R. Bruce, Burns, James A., & Young, Robert R. (1994) Postglacial alluvial terraces and an incorporated bison skeleton, Ghostpine Creek, southern Alberta. *Canadian Journal of Earth Sciences*, vol. 31, no. 10, pp. 1501–9.

Sandweiss, Daniel H., Kelley, Alice R., Belknap, Daniel F., et al. (2010) GPR identification of an early monument at Los Morteros in the Peruvian coastal desert. *Quaternary Research*, vol. 73, no. 3, pp. 439–48.

Sternberg, Ben K. & McGill, James W. (1995) Archaeology studies in southern Arizona using ground penetrating radar. *Journal of Applied Geophysics*, vol. 33, no. 1, pp. 209–25.

Tolksdorf, Johann Friedrich, Turner, Falko, Kaiser, Knut et al. (2013) Multiproxy analyses of stratigraphy and palaeoenvironment of the late Palaeolithic Grabow floodplain site, northern Germany. *Geoarchaeology*, vol. 28, no. 1, pp. 50–65.

Valdés, Juan Antonio & Kaplan, Jonathan (2000) Ground-penetrating radar at the Maya site of Kaminaljuyu, Guatemala. *Journal of Field Archaeology*, vol. 27, no. 3, pp. 329–42.

Waters, Michael R. (1992) *Principles of Geoarchaeology, A North American Perspective*. The University of Arizona Press, Tucson, Arizona.

Waters, M.R. (2008) Alluvial chronologies and archaeology of the Gila River drainage basin, Arizona. *Geomorphology*, vol. 101, pp. 332–41.

5

Soils Horizons, Peat Bogs, and Swamps

Abstract: Soil horizons within stratigraphic sequences document periods of stability on which people lived and performed other activities. These important geoarchaeological units can be identified in GPR reflection profiles and traced across ancient landscapes to provide important spatial information about now buried environments. Swamps and peat bogs are fresh water environments where radar energy can often penetrate deeply to produce images of thick sequences of clastic and organic material. Changes in these wetland environments can be identified with GPR as they often began as lakes and springs, filled in with organic matter and were finally transformed into swamps and bogs. Human habitations visible with GPR within these environments can consist of platforms for living and work activities.

Keywords: soil, peat, swamps, bogs, human habitation, anthropogenic layers

Introduction

Soil units play an important role in the understanding of past landscapes as they provide evidence of relative environmental stability for some period of time (Birkeland 1999). It was often on those stable surfaces that people lived and performed many other activities leaving an abundant archaeological record. When soils form on stable landscapes they produce a continuous blanket across all previously deposited units and when these units can be identified and then mapped with GPR, the surface on which people lived can be understood in three dimensions. Soil units commonly display three (and sometimes more) distinct subunits with the uppermost horizon an A unit that forms by organic matter accumulation. Below the A, the B horizon grows downward over time due to the accumulation of clay, iron oxides, and carbonate and by the weathering of the parent material (often composed of bedrock or previously deposited sediment). The lowest horizon is partially weathered parent material, termed the C zone (Birkeland 1999).

There are many variables in soil types and constituents that are partially dependent on the environment within which they formed, the type of parent material, and the amount of time the landscape was stable and allowed them to form (Birkeland 1999). Other variables that can potentially be quantified in soil development, beyond just their thicknesses, are the amount of biological mixing that occurred and the types and amounts of organic and mineral constituents added to the units during their formation

Ground-penetrating Radar for Geoarchaeology, First Edition. Lawrence B. Conyers.
© 2016 John Wiley & Sons, Ltd. Published 2016 by John Wiley & Sons, Ltd.

(Doolittle and Butnor 2009). All these activities and resulting constituent properties of soil units influence the velocity of radar waves that propagate through them, due to variations in the way water is retained and distributed. At interfaces between soil horizons and sedimentary layers, and also from physical changes within the soils themselves, boundaries between all these horizons can reflect radar energy to be recorded and interpreted.

Another factor in soils to take into consideration is that surface soil types and their constituents affect the way energy leaves the antenna and couples with the ground during transmission and also its propagation depth (Doolittle et al. 2007). An analysis of surface soils can often be used as a way to help understand attenuation depth and potential resolution for GPR in different areas. The largest factor in surface soils that affects radar propagation is how electrically conductive these units are (Conyers 2013, p. 75; Neal 2004; Shainberg et al. 1980). As the amount of salt and the percentage of certain mineralogies of clay increase, the **cation exchange capacity (CEC)** of a soil will also increase, making it more electrically conductive. Soils with high CEC are those with smectite and vermiculite clay types such as montmorillonite and bentonite (Saarenketo 1998). Soils, or any sediments with these types of clays, especially when wet, will attenuate radar energy at a very shallow depth. The low CEC clays are kaolinite, gibbsite, and other halloysitic clay types, which allow radar penetration with little attenuation (Doolittle et al. 2007). Usually soils that form in tropical areas with a large amount of rainfall, or in other environments with a good deal of groundwater leaching, will contain the low CEC clays, and are excellent media for GPR energy penetration. Calcium carbonate accumulation in a B soil horizon, in what are termed caliche or "hardpan" units, implies poor water leaching and therefore the likely preservation of salts or high CEC clays near the surface—and therefore poor radar energy penetration. Ground containing any type of salt can also be a poor environment for GPR (Conyers 2012, p. 95).

The GPR method has been used in agricultural studies to show root biomass, soil architecture, tree root propagation, and water retention properties (Doolittle and Butnor 2009). These types of properties can potentially be incorporated in geoarchaeological studies, but to date this has not been attempted in any comprehensive way. Theoretically buried soil horizons that are encased within sediment packages, and which were farmed in the past, could be analyzed for these same properties important for agricultural people in the past. As the GPR method collects data on depth and a great deal of amplitude and frequency information of reflected energy from buried soil horizons, these variables could possibly be calibrated for agricultural productivity variables. In this way the factors that were important in ancient agricultural activities, their productivity, and possibly other interesting information of this sort could be interpreted over large areas.

Soil Horizons

In 1926 excavations took place near Folsom, New Mexico, at what turned out to be the first documented site in the Western Hemisphere containing the association of prehistoric artifacts and now-extinct Pleistocene megafauna of the species *Bison antiquus* (Meltzer et al. 2002). Excavated by the Colorado Museum of Natural History (now Denver Museum of Nature and Science), the Smithsonian Institution, and the American Museum of Natural History, the site contained remains of between 30 and 50 bison that were killed in a gully that was rapidly filled with sediment, preserving the bones and archaeological remains from about 10,000 BP. On the south bank of Wild Horse Arroyo the partially backfilled remains of the old excavations leave a scar on the landscape still visible where the main bone bed was excavated. The north bank appears to contain alluvium that filled this fluvial channel, which has been tested by only a few excavations and core holes.

Much of this channel-fill material, which is up to 4–5 m thick, began to be deposited about 12,400 BP (Meltzer et al. 2002) with fluvial and colluvial sediment derived from nearby outcrops of the Cretaceous-age shale. The filling continued until at least 4400 years ago or perhaps later, with erosion and the exposure of these sediments along the gully margins only occurring in recent times. The bone bed that was excavated is located in a small tributary gully to the south of the main watercourse. There is some evidence that bones may still be preserved in the north bank in unexcavated alluvial fill where GPR tests were performed.

To test the ability of GPR to produce images of the geological layers that filled in this small valley, a number of reflection profiles were collected, where the reflections could be tied directly to the sediment and soil units visible in the cut banks (**Figure 5.1**). It was immediately apparent that the 400 MHz antennae were only capable of transmitting radar waves to about 1 m in this sediment. Any GPR imaging of the deeply buried bone bed that was predicted to be about 4 m deep in the deepest part of the paleo-channel was therefore not possible. However, one well-developed soil unit of unknown age is exposed along the bank of the gully, and the GPR reflection profiles could be "tied" directly to it for correlation. These soil horizons are within the upper 1 m of the channel fill package.

The top of a buried A soil is about 25 cm below the ground surface, overlain by a fluvial sand and capped by a surface A soil (**Figure 5.1**). The contact between the top of the buried soil (Ab in **Figure 5.1**) with the underlying sandy unit (C in **Figure 5.1**) readily produces a high-amplitude reflection. The same type of planar reflection was generated at the contact of the base of the surface soil with the fluvial sand. A third reflection was generated at the base of the buried soil where there is a clay-rich calcium carbonate-impregnated B zone (Btkb horizon in **Figure 5.1**). These soil horizons and their contacts with the fluvial sand are interfaces parallel to the present ground surface, and all generated good radar reflections. The planar reflection surfaces exhibit little variation in depth or reflectivity away from the outcrop (**Figure 5.1**). However, all these reflections abruptly end about 4 m away from the outcrop, where there are very different reflections from presumably different stratigraphic units that have varying thicknesses and amplitudes. This change in GPR reflection is exactly the distance away from the edge of the exposure where the paleo-arroyo was inferred by core hole data (Meltzer et al. 2002). It is likely that this vertical interface visible in the GPR profile is displaying the contact between valley fill units and soil horizons that had formed on the surrounding exposed bedrock. The buried soil visible in the outcrop (Ab-Btkb) should merge with those soils that were forming on the gully edge at the time it was buried, but this continuation of the reflection is not immediately apparent in the reflection profile. Perhaps the soil layer that formed on the adjacent bedrock has a different composition and reflects energy differently than the more well-developed unit in the valley bottom, and in the GPR profile this amplitude change appears as a lateral discontinuity.

In southern Arizona along the bank of a river where a buried prehistoric irrigation canal and associated soil was discovered (**Figure 4.9**) two soil units are exposed that appear to correlate to the nearby soil surface presumably used for agriculture about AD 1200 (**Figure 5.2**). Both contain poorly developed A horizons with very weak B horizons. As they occur at the same elevation as the unit visible with GPR in the golf course about 100 m to the south (**Figure 4.9**), one or both are hypothesized to be correlative to the hypothesized prehistoric maize field soil.

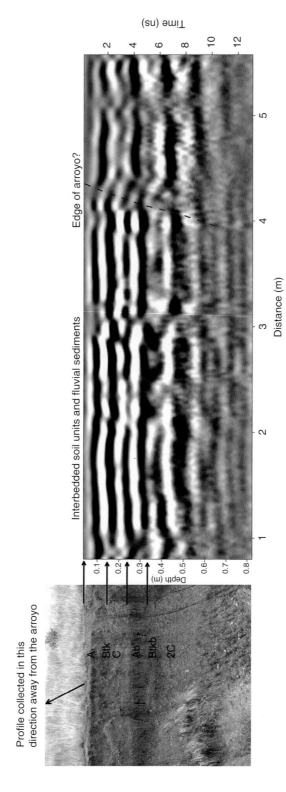

Figure 5.1 Correlation of surface and buried soil horizons in an alluvial fill sequence from northern New Mexico. Individual soil horizons each generate high-amplitude reflections in this 400 MHz profile.

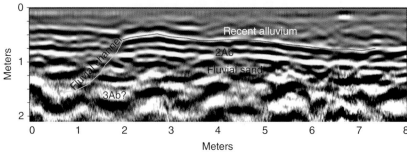

Figure 5.2 A 900 MHz reflection profile showing reflections produced from alluvial and soil units along a braided river in southern Arizona. The soil horizons produce distinct reflections while the fluvial sands are largely devoid of reflections as they have no interfaces to reflect energy.

This exposure is quite complex with two buried soil units overlaying a basal fluvial sandy gravel bed visible at the base of the exposure (the lowest fluvial sand in **Figure 5.2**). Overlying the basal fluvial sand is a sequence of alternating thin fluvial channels where fluvial deposition was interrupted by periods of stability where soils formed. It is likely that during the periods of relative stability the braided channels of this dynamic river eroded downward, leaving small terraces along the valley margins that were not subject to regular flooding, where these soils could form for at least a few tens of years. It was along these small terraces, just above the active floodplain, that prehistoric people constructed irrigation canals and farmed the developing soils on the small terrace treads. The area was still flood prone and even these raised terrace surfaces experienced periodic sand deposition as the main channel filled with sediment and smaller river channels spilled over on to the terrace surface, covering the agricultural fields with sand. The river again degraded into the main channels, leaving this area stable for some time when soils again formed. This last period of stability is labeled the 2Ab horizon in **Figure 5.2**. Again, after some time of stability, a fluvial channel about 2–3 m in width cut through the previously developed and deposited units, probably from a tributary stream that flowed into the main river channel, still visible in the vicinity. This channel incised downward through all the previously deposited units, which was then filled with dark redeposited sediment consisting of a mixture of organic-rich A soil and sand.

The main river valley again aggraded, covering the floodplain and the small adjacent terrace surfaces where soil was forming. This fluvial sediment filled up to the top of the present exposure in **Figure 5.2**.

The uppermost unit in this sequence consists of a thin cross-laminated sandy gravel bed containing plastic and metal artifacts presumably deposited in the last 50 years (Pearthree and Baker 1987). Periodic torrential floods since 1982 led to a sustained period of degradation where the river channel incised many meters into the previously deposited sediments and then partially refilled to its present level (just at the bottom of the photo in **Figure 5.2**). These types of dramatic aggradation, degradation, and intervening short periods of stability have been well documented in this river valley and elsewhere in southern Arizona (Haynes and Huckell 1986; O'Mack et al. 2004). Dynamic changes along this watercourse would have made prehistoric agriculture tenuous and a constant battle with the forces of the river. However, there were decade- or century-long periods of environmental stability on the small terraces above the floodplain where people succeeded in growing crops, as can be seen in the soil units.

A 900 MHz reflection profile was collected about 50 cm away from the edge of the river bank where these sediment and soil units are exposed (**Figure 5.2**). Velocity was calculated using hyperbola fitting and the units visible in the reflection profile were directly correlated to those in the outcrop. The uppermost sediment package, consisting of the recently deposited sandy gravel, is displayed as a mostly homogeneous layer in the GPR profile. It has only a few hints of the cross-laminated layers visible in the outcrop, none of which reflected radar waves. The basal contact of the recent alluvium unit with the underlying buried soil unit (labeled 2Ab in **Figure 5.2**) produced a distinct planar reflection. The reflection from this interface can be traced laterally, where it was eroded by the fluvial channel that flowed down a small tributary into the main floodplain. This tributary channel is visible as a sloping interface with very different reflection layers that filled this erosional feature (**Figure 5.2**).

In the preserved area to the right of the fluvial channel the contact between the 2Ab soil unit and the underlying fluvial sand is also evident by a high-amplitude planar reflection generated from the contact of the soil unit and the poorly layered underlying fluvial sand (**Figure 5.2**). The fluvial sand also generated a few very low-amplitude jumbled radar reflections, generated from its cross-laminations. The contact between the 2Ab soil and the lowest fluvial sand in this sequence is not readily visible, as radar energy was attenuated below about 1.5 m depth. A 400 MHz profile was also collected along this exposure in the hope of transmitting energy deeper in the ground, but energy of that frequency was attenuated at exactly the same depth as the 900 MHz, showing that this ground is electrically conductive with attenuation below 1.5 m irrespective of the frequency. The 400 MHz reflection profile also had much less definition of these buried soils and sediment units and was not useful in this stratigraphic study.

This soil and sediment test (**Figure 5.2**), much like the one from Folsom, New Mexico (**Figure 5.1**), demonstrates that GPR is very useful for defining buried soil horizons, which can easily be discriminated from interbedded sediment layers. The edges of sediment and soil packages along valley margins as well as those cut by fluvial channels can also be identified by poorly defined vertical or sloping interfaces with very different reflection features on either side. It is the difference in soil types of sediment constituents on either side of those interfaces that are key to identifying the more vertical interfaces of this sort in GPR reflection profiles.

In coastal Portugal, some quite ancient buried soil units are exposed within an aeolian sand sequence (**Figure 5.3**). Here, the soil of interest is a late Pleistocene–early Holocene horizon with thick A and E horizons on top of a well-developed clay-rich Bt unit

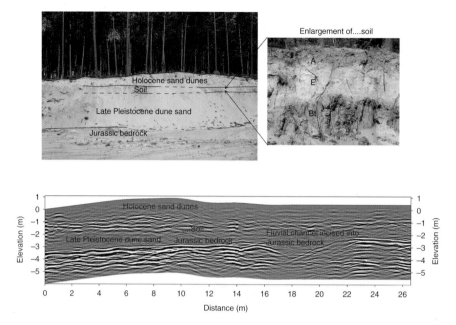

Figure 5.3 Buried late-Pleistocene soil with distinct E and B horizons within an aeolian sand package from western Portugal. The 270 MHz reflection profile displays lateral variations within this buried soil. The resolution is not high enough to differentiate each of the soil subunits.

(Birkeland 1999). The A horizon has been depleted in organic materials over the thousands of years as it has been subject to weathering and leaching from downward water movement. There was also a good deal of leaching by groundwater movement during the time this soil was developing, creating the leached E horizon consisting of almost pure quartz sand (**Figure 5.3**). Any clay, salts or other compounds that might have been brought into this area by the wind, or weathered from the original parent material, have been translocated into the lowest horizon in this profile, the Bt. Sometime in the mid-Holocene this soil was buried by encroaching sand dunes and became a relict soil (Birkeland 1999), now buried by 1–2 m of Holocene aeolian sand. The parent material for this late Pleistocene–early Holocene soil is a late Pleistocene aeolian unit, which sits directly on the Jurassic age bedrock (**Figure 5.3**).

A 270 MHz reflection profile was collected along the top of this exposure. Excellent depth penetration occurred with good radar reflections received from the Jurassic bedrock surface at 3–4 m depth (**Figure 5.3**). A fluvial channel incised into the Jurassic bedrock is visible in the GPR reflection profile, much like those shown in **Figure 4.3**, where they were studied about 250 m to the east of this location. The late Pleistocene dune sand is essentially invisible to radar waves, producing no reflections from this quartz-rich aeolian deposit. The overlying soil unit is visible as a discontinuous high-amplitude planar reflection (**Figure 5.3**). What causes this reflection surface to exhibit laterally varying amplitudes is not known, but could be differences in the water-holding capability of the Bt horizon. While no detailed analysis of these soil units was conducted, the ability of clay units to retain water in their molecular structure would allow the Bt horizon to contrast a great deal with the bounding sands producing these reflections. The permeable sand units would have easily been drained of moisture in the summer dry season while the Bt retained water, creating the velocity contrast necessary to generate high-amplitude radar reflections (Conyers 2012, p. 34).

With topographic corrections of many reflection profiles collected in a grid, the topography of the late Pleistocene–early Holocene stable living surface denoted by the buried soil, could be mapped in three dimensions with GPR. If artifacts or other archaeological features were known to exist on and within this soil (they were not identified here), the mapping of this soil unit could allow for ancient landscape analysis across a large area.

Swamps and Peat Bogs

Some of the first applications for GPR were to measure depth of permafrost and peat deposits for civil engineering purposes (Ulriksen 1992). These early studies were conducted in the winter when the ground surface was frozen and often snow covered, which allowed antennae and people to more easily move across otherwise boggy ground. In different areas of the world these wetland areas are termed marshes, swamps, mires, moors, and peat bogs, all of which are different in some respects but common in that they are very high in organic matter and wet or saturated much of the year.

Since those early studies in permafrost areas, many applications of GPR have been used in these environments to map stratigraphic layers within the organic-rich units (Nobes 1992; Persico et al. 2010; Ruffell and McKinley 2014) and archaeological materials within them (Damiata et al. 2013; Utsi 2004). Researchers consistently describe good radar energy penetration, sometimes up to 7 m deep (Ruffell et al. 2004), but always consistently low radar wave velocity. The RDP of fresh water is 80, and with the usually saturated or partially saturated ground in bog areas, the RDP for these materials as a whole can be as high as 70, making radar travel rates only about 20% of those waves moving in air (Table 2.1). As fresh water is not electrically conductive, the depth of energy penetration is still excellent in this type of ground and the shorter wavelengths of downloaded energy (Table 2.1) will allow good resolution of interfaces and other features (Conyers 2013, p. 64). Both depth and resolution with GPR in peat bogs make this an almost uniformly good environment for GPR studies.

In southwestern Scotland there are many peat bogs that are regularly harvested for fuel, and around which people have lived for centuries (Utsi 2004). One of these, located very near sea level, is locally termed a "moss," around and within which many interesting artifacts were discovered in the 19th century. Circular wattle structures, stone tools, and a container of "bog butter" (wooden cask with dairy products or animal fat placed in the bog for storage) were found nearby (Christianson 1881). The most interesting artifact was a life-sized wooden figure of a female, which was placed face down in the bog, pinned to the bottom with bent sticks in much the same way bog bodies were interred in many other areas of the British Isles and northern Europe. All were dated to the about 3000 BP or earlier (Utsi 2004). It was speculated that there might be Mesolithic sites below the peat layers that were perhaps occupied when the bog was still a lake, which could potentially be found with GPR.

A grid 60×50 m in size was collected over the surface of the bog with 50 MHz antennae (Utsi 2004). The profiles show that well-defined reflections were recorded to about 5 m in this boggy ground, with an RDP of 73. A distinct basal horizon composed of unknown but well-layered sediments dips into the middle of the peat bog. This interface is likely the ancient ground surface prior to the formation of an initial fresh water lake in this area (**Figure 5.4**). The horizontal layers that filled this depression reflect very little energy, which is expected with thinly laminated lacustrine sediments and a low-frequency 50 MHz antenna capable of transmitting waves of about 70 cm wavelength in this material.

The reflection profile shows that by the time the lake had filled with about 2–3 m of sediment, platforms were built on its shore, which are visible as high-amplitude raised features (**Figure 5.4**). Four of them were found with GPR in this peat bog, and one was cored to recover a clay layer associated with wooden artifacts and sandstone and quartz clasts, which were presumably used as paving on the top of a platform.

The GPR profile shows that the two platforms shown here are composed of multiple layers that were likely renovated and raised as much as 1.5 m over time as the lake or surrounding bog water level rose. The reflective nature of the sediment surrounding the platforms shows a distinct discontinuity at about 2 m depth, which may be the transition from the initially deposited interbedded lake and organic-rich beds to almost pure peat at the top. This type of depositional transition was common in this part of Scotland as deforestation occurred in the area surrounding lakes, which led to greater runoff and more waterlogged ground in lower areas over time. In this low area lake sediments gave way to peat formation during the last 2000 years or so, until these platform features were totally buried in organic matter.

At high altitude in the Rocky Mountains of Colorado a number of small wet areas are present in valley bottoms, locally termed fens or bogs. These small wetland areas are ground-water fed and consistently water saturated, with only a minor amount of drying in the summer. It is in similar environments, but usually more extensive, where bog bodies and other interesting archaeological features have been discovered in Europe (Menotti and O'Sullivan 2013; van der Sanden 2013). A number of GPR profiles were collected over this bog in the winter when it was frozen and covered with a thick snow layer. One profile collected with 400 MHz antennas displays this small bog, which is about 10 m in diameter (**Figure 5.5**). The profile was corrected for depth using a RDP of 65, which accurately shows the depth of the bog feature, but makes the surface snow and ice layers appear much too thin (as frozen water has a RDP of between 3 and 4 (Conyers 2013, p. 50), and radar velocities within that surface unit were very high.

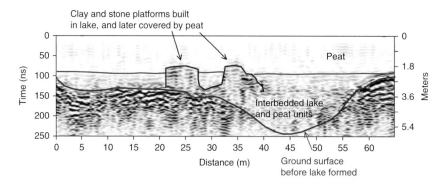

Figure 5.4 A 50 MHz center reflection profile collected across a peat bog in southwestern Scotland displaying the high-amplitude reflections deposited in lacustrine sediments before the lake and subsequent peat bog was formed. Raised platforms on the margin of the bog, projecting out into the center of the wetland, were work areas and possible habitation structures from late Mesolithic times. Data from Erica Utsi.

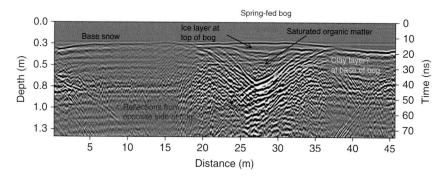

Figure 5.5 A 400 MHz reflection profile collected on snow across a small peat bog in the Rocky Mountains, Colorado. This small fen shows a basal clay surface, which was likely deposited in a spring before it filled with organic material. From Sarah Lowry.

The ice layer just below the snow shows thickening in the middle of the bog with thinning on its edges. The saturated organic matter within the bog displays very few high-amplitude reflections, as it is composed primarily of peat with no stratigraphic interbeds (**Figure 5.5**). The base of the bog at about 50 cm below the ground surface is a highly reflective unit, which might be a clay layer. This presumed clay unit may have been deposited as the bog was just forming, as spring deposit, and over time organic matter filled in the spring when environment changed to a bog. The clay continues to act as an aquitard and therefore water is retained in this area even during summer months. As this clay unit is highly reflective, it produces a high-amplitude reflection in much the way that irrigation canals do, with energy from the surface antennae moving at an oblique angle to the bog edges, recorded here as one half of a "bow-tie" reflection feature (**Figure 3.8**).

References

Birkeland, Peter (1999) *Soils and Geomorphology*, 3rd Edition. Oxford University Press, New York.

Christianson Robert (1881) On an ancient wooden image, found in November last at Ballachulish Peat Moss. *Proceedings of the Society of Antiquaries of Scotland*, vol. 15, pp. 158–78.

Conyers, Lawrence B. (2012) *Interpreting Ground-penetrating Radar for Archaeology*. Left Coast Press, Walnut Creek, California.

Conyers, Lawrence B. (2013) *Ground-penetrating Radar for Archaeology*, 3rd Edition. Altamira Press, Rowman and Littlefield Publishers, Lantham, Maryland.

Damiata, Brian N., Steinberg, John M., Boldender, Douglas J., & Zoëga, Guðný (2013) Imaging skeletal remains with ground-penetrating radar: comparative results over two graves from Viking Age and Medieval churchyards on the Stóra-Seyla farm, northern Iceland. *Journal of Archaeological Science*, vol. 40, no. 1, pp. 268–78.

Doolittle, James A. & Butnor, John R. (2009) Soils, peatlands, and biomonitoring. In: Harry M. Jol (ed.) *Ground Penetrating Radar Theory and Applications*, pp. 179–202. Elsevier, Amsterdam.

Doolittle, James A., Minzenmayer, Fred E., Waltman, Sharon W., Benham, Ellis C., et al. (2007) Ground-penetrating radar soil suitability map of the conterminous United States. *Geoderma*, vol. 141, no. 3, pp. 416–21.

Haynes, C. Vance & Huckell, Bruce B. (1986) *Sedimentary Successions of the Prehistoric Santa Cruz River, Tucson, Arizona*. Open-file Report 86–15, Arizona Bureau of Geology and Mineral Technology, Tucson, Arizona.

Meltzer, David J., Todd, Lawrence C., & Holliday, Vance T. (2002) The Folsom (Paleoindian) type site: past investigations, current studies. *American Antiquity*, vol. 67, no. 1, pp. 5–36.

Menotti, Francesco & O'Sullivan, Aidan, eds. (2013) *The Oxford Handbook of Wetland Archaeology*. Oxford University Press, Oxford.

Neal, Adrian (2004) Ground-penetrating radar and its use in sedimentology: principles, problems and progress. *Earth-Science Reviews,* vol. 66, no. 3, pp. 261–330.

Nobes, David C. (1992) Discussion of the utility of conductivity surveying and resistivity sounding in evaluating sand and gravel deposits and mapping drift sequences in northeast Scotland. *Engineering Geology*, vol. 33, no. 2, pp. 151–2.

O'Mack, Scott, Thompson, Scott, & Klucas, Eric Eugene (2004) *Little River: An Overview of Cultural Resources for the Rio Antiguo Feasibility Study, Pima County, Arizona*. Statistical Research Inc. Technical Series No. 82, Statistical Research, Tucson, Arizona.

Pearthree, Marie Slezak & Baker, Victor R. (1987) *Channel Change Along the Rillito Creek System of Southeastern Arizona 1941 Through 1983*. Special Paper No. 6, Arizona Bureau of Geology and Mineral Technology, Geological Survey Branch of Arizona, Tucson, Arizona.

Persico, Raffaele, Soldovieri, Francesco, & Utsi, Erica (2010) Microwave tomography for processing of GPR data at Ballachulish. *Journal of Geophysics and Engineering*, vol. 7, no. 2, pp. 164–77.

Ruffell, Alastair & McKinley, Jennifer (2014) Forensic geomorphology. *Geomorphology*, vol. 206, pp. 14–22.

Ruffell, Alastair, Geraghty, Louise, Brown, Colin & Barton, Kevin (2004) Ground-penetrating radar facies as an aid to sequence stratigraphic analysis: application to the archaeology of Clonmacnoise Castle, Ireland. *Archaeological Prospection*, vol. 11, no. 4, pp. 247–62.

Saarenketo, Timo (1998) Electrical properties of water in clay and silty soils. *Journal of Applied Geophysics*, vol. 40, pp. 73–88.

van der Sanden, Wijnand (2013) Bog bodies: underwater burials, sacrifices and executions. In: Francesco Menotti & Aidan O'Sullivan (eds.) *The Oxford Handbook of Wetland Archaeology*, pp. 401–32. Oxford University Press, Oxford.

Shainberg, Isaac, Rhoades, J.D., & Prather, R.J. (1980) Effect of exchangeable sodium percentage, cation exchange capacity, and soil solution concentration on soil electrical conductivity. *Soil Science Society of America Journal*, vol. 44, no. 3, pp. 469–73.

Ulriksen, C.P.F. (1992) Multistatic radar system-MRS. In: Pauli Hanninen & Sini Autio (eds.) *Proceedings of the Fourth International Conference on Ground-Penetrating Radar: June 8–13, Rovaniemi, Finland*, pp. 57–63. Geological Survey of Finland Special Paper 16, Helsinki.

Utsi, Erica (2004) Ground-penetrating radar time-slices from North Ballachulish Moss. *Archaeological Prospection*, vol. 11, no. 2, pp. 65–75.

6 Beaches, Sand Dunes, and other Coastal Environments

Abstract: Coastal environments can be geologically complex due to high-energy agents that move sediments by wind, currents, waves, and storm activity. Changes in relative sea level affecting the locations of human habitations over time produce additional complexity. Beaches are often visible with GPR as non-reflective blankets of sand with a few poorly defined seaward dipping reflections. Near-shore sand dunes have often been locations of habitation and other activities that can be complex as these mobile blankets of sediment have migrated over time, burying some sites while others are eroded, destroyed or re-worked. The erosional contacts that were produced from all of these environmental shifts can be identified in GPR profiles and non-disturbed areas where people lived and worked within them mapped.

Keywords: aeolian, dunes, beaches, sea level change, erosional contacts, burials, work areas

Introduction

Coastal environments have long been the locations of human activities including hunting and gathering, fishing and exploitation of marine resources, habitation structures and trading activities (Waters 1992). There are numerous depositional environments in these areas that can be identified geologically and studied with GPR, which range from beaches, near-shore sand dunes, barrier islands and back-barrier areas, to name just a few (Pearl and Sauck 2014). These were areas where sediment was transported by marine actions such as waves, longshore drift and tides, but also modified by wind and fluvial processes. They can be especially complex environments as coastlines can be rapidly altered by dramatic storm activity and also affected during longer time periods by sediment accumulation, erosion, and sea level changes.

Sea level variations had a significant impact on human activities in the long term, as some near-shore environments important to people were drowned by rising water levels, or conversely important ecosystems such as bays, swamps, and estuaries were drained of water by lowering sea level. While we often associate sea level changes with major climatic shifts, such as the end of the last Ice Age, smaller relative sea level changes also affected coastal environments as a function of subsidence, tectonic uplift or downwarp and the progradation of shorelines accompanied by sediment influx (Douglas 1997; Peltier 2002).

Ground-penetrating Radar for Geoarchaeology, First Edition. Lawrence B. Conyers.
© 2016 John Wiley & Sons, Ltd. Published 2016 by John Wiley & Sons, Ltd.

These dynamic coastal areas containing archaeological sites are often readily buried and preserved by deposition of sediments, but also destroyed by slow erosion or the dramatic destructive forces of storms. Environments of deposition, people's use of ancient environments, and their material record preserved within sediment packages can be readily identified using GPR images (Bristow and Pucillo 2006). Geoarchaeological investigations integrated with GPR can potentially identify even minor ocean transgressions and regressions along a coastline (Møller and Anthony 2003; Moore et al. 2004) and coastal landscape changes reconstructed (Neal and Roberts 2000).

Beaches

In the eastern Mediterranean Sea GPR was used to explore for possible ancient activities related to the beaching of ships at the site of Ashkelon, Israel (**Figure 6.1**). This huge site has been excavated for decades, and there is abundant evidence for the trade of goods that moved by ship having taken place here, but no port or other facilities related to that trade have been found (Master 2003). Offshore shipwrecks that provide additional evidence for shipping have been found nearby (Ballard et al. 2002) but no evidence of how items were loaded and unloaded adjacent to the ancient city.

Many reflection profiles were collected on the narrow beach just seaward of the site in the search for port facilities (**Figure 6.1**). It was evident from viewing the exposures in the sea cliffs that there has been a good deal of erosion along this coastline over time as architectural materials from many different periods of occupation are exposed in the sea cliff sediments and archaeological materials are visible on the beach that have been eroded by wave action. Large architectural blocks of buildings that were once located on presumably stable ground surface many tens of meters above sea level are now visible in the shallow water tens of meters offshore (**Figure 6.1**). This suggests a great deal of

Figure 6.1 Location on the coast of Israel where 400 MHz reflection profiles were collected in the search for coastal trade facilities.

coastline erosion over just a few millinea. While the possibility of finding docks, quays, or other facilities for shipping seemed unlikely, as they would be present far offshore due to coastline erosion, reflection profiles were collected on the beach and display some interesting near-shore features.

Exposed in places along the beach cliff and inland is a well-cemented formation of late Pleistocene–early Holocene aeolian sand deposited during a time that sea level was much lower than today (Moshier et al. 2011). It is visible along the base of the seacliff and in small outcrops in the beach sand and is locally referred to as "beach stone" (**Figure 6.1**). This bedrock unit is composed of mostly quartz sand with clay and carbonate cement. Just a few hundred meters to the east of the beach this unit is exposed at a higher elevation as resistant sandstone ridge composed of well-developed aeolian cross bedding, locally called *kurkar*. As humans began to occupy this area in the Neolithic, they likely built their first structures directly on this resistant bedrock unit at a time when the coastline was some distance to the west and the initial construction occurred on small topographic rises along the beach. Many thousands of years of anthropogenic tell sediments have been deposited during the last 6000 years or so, creating the voluminous anthropogenic deposit, which is the site of Ashkelon.

The 400 MHz reflection profiles collected on the Ashkelon beach display high-amplitude planar reflections from the *kurkar* aeolian bedrock, with the present day beach sand exhibiting much lower amplitudes (**Figure 6.2**). The bedrock is highly reflective because its layers have been cemented differently, each having a different permeability and retaining water in varying amounts, producing velocity discontinuities to reflect radar waves. Recently deposited beach sand is well winnowed and uncemented, and while displaying some weak layering, the reflections are much lower in amplitude. These beach layers dip gently seaward with all reflections ceasing in the swash zone where the high electrical conductivity salt water attenuates all radar energy (**Figure 6.2**). As expected, no archaeological features that might be related to ancient trade were found with GPR.

The Island of Guam in the western Pacific Ocean is an interesting case study in sea level change and human adaptation to those changes during the last 3000–3500 years (Carson 2011). The archaeological record indicates that people first reached this island about 3500–3000 years ago when relative sea level was at a high stand, about 1.5–2 m above its present level. At that time, there were extensive mangrove swamps and back-beach lagoons and marshes, protected from the ocean by sandy beaches. These resource-rich environments were exploited by the first immigrant people and their pottery and other artifacts are found associated with these areas as well as along the beaches just above what were the wave-impacted zones during that high-stand sea level.

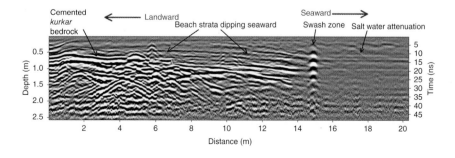

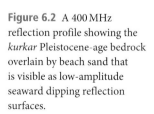

Figure 6.2 A 400 MHz reflection profile showing the *kurkar* Pleistocene-age bedrock overlain by beach sand that is visible as low-amplitude seaward dipping reflection surfaces.

In the Guam area, and much of the Marianas Island chain, sea level then began to recede during a period of fore-arc tectonic uplift (Dickinson 2000, 2003) and the shoreline and accompanying beach sands prograded seaward. As sea level dropped, mangrove swamps were drained and the shallow coral reefs that were growing very near sea level were exposed. Wave-cut notches, which formed as waves eroded the previously deposited limestone, were exposed and soon covered by sand dunes and sheet-wash deposits derived from the higher inland areas.

Today, the sedimentary record of beach sediments and the time-correlative wave-cut notches that formed at the time of high stand sea level are preserved below thin aeolian deposits about 2 m above the present sea level. A 400 MHz GPR profile was collected on the north shore of Guam near Ritidian, with the transect beginning in the salt water and ending inland near stabilized coastal dunes. The profile shows one wave-cut notch incised into earlier deposits that was formed probably as sea level rose. A second notch was cut at the sea level high-stand (**Figure 6.3**). The highest elevation notch is at about 1.5 m above present sea level, which is the approximate high level of that sea level maximum (Dickinson 2003). Just landward from that high-stand notch artifacts from the first inhabitants of Guam have been found elsewhere on Guam, dating to between 3000 and 2500 BP (Carson 2011).

The GPR profile also displays thin non-reflective sediment units seaward of the wave-cut notches, which are typical of well-sorted beach sand. Other reflection profiles collected on areas where the beach is wider show two units of beach sand both with seaward dipping strata, bounded at the base by a buried soil or some other type of erosional surface (**Figure 6.4**). It is unlikely that any artifacts would be found in these units except for abraded and out of place objects, as these units were deposited under high wave energy.

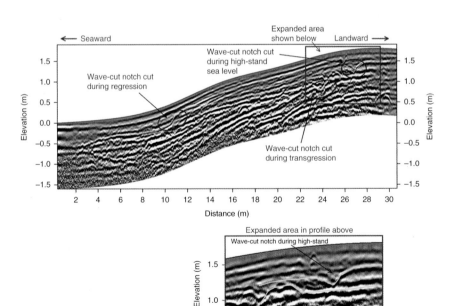

Figure 6.3 A topographically adjusted 400 MHz reflection profile showing wave-cut notches and seaward beach sand produced during transgressive and regressive events on the north coast of Guam. Data from Michael Desilets.

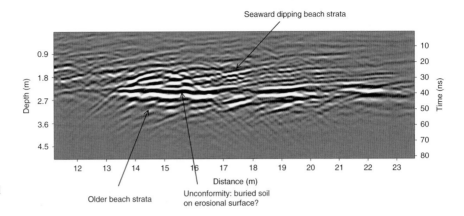

Figure 6.4 Seaward dipping beach strata sitting on an unconformable lower surface from Guam. Data from Michael Desilets.

Erosion Features along Coasts

Wave-dominated coastlines, or those that are periodically impacted by intense storms, typically have erosional features similar to those of Guam's wave-cut notches. On the island of Key West, Florida, the local bedrock (locally called "hardpan") is the Miami Oolite, a well-indurated Pleistocene limestone formation (Muhs et al. 2011). Periodic hurricanes have eroded portions of the coastline, and this shallow resistant bedrock unit has eroded and retreated in places landward producing small beach-cliffs. Post-storm deposition of beach and aeolian sand then buried this erosional scarp. It was important to differentiate these units in a GPR survey conducted here in the search for graves within the beach and aeolian sand.

This search for historic graves was conducted to find burials of African slaves and pirates (Conyers 2012, p. 147) in what is today a Key West city park. Many grids of GPR data were collected along the south coast in preparation for park redevelopment and road widening. This area in the 18th and 19th centuries was used as an informal burial ground for those not affluent enough to be buried in the city cemetery, or others of low standing in society. The area surveyed on the south shore had very narrow beaches in the past and it was in the broad sand dunes just inland from the beaches that burials took place. The beach and aeolian sand was periodically washed away by hurricanes and other storms, and there are historical accounts of skeletons found on the beach after these violent events. Today, sand is rapidly replaced after storms by the city to make sure one of the larger beaches on the island can be enjoyed by residents and tourists. Much of the old sand dune area adjacent to the beach has also been recently stabilized by planting lawns in a large park and the construction of bicycle paths and roads. Historic documents and oral histories tell of this area being the burial location of pirates who were captured at sea and then put to death and buried in these dunes (Conyers 2012, p. 147). During World War II there were military barracks constructed in the survey area and written accounts of pipe-laying construction workers commented on the abundance of human bones uncovered during those operations.

The GPR reflection profiles collected in the area of the presumed burial ground show good energy penetration to about 1–1.5 m in this find-grained carbonate sand using 400 MHz antennas (**Figure 6.5**). The most prominent reflection visible was generated from a very high-amplitude planar reflection, known from probes and cores to be the oolite bedrock, or "hardpan." This layer is periodically exposed during wave erosion during storms, and then covered over again by the city maintenance workers. Everyone who excavates in Key West for a living knows how difficult it is to penetrate this bedrock unit, and it is avoided as much as possible.

With this information in hand, the areas where the bedrock was buried deeper than 80 cm, or had been completely removed by erosion and the area backfilled with beach and aeolian sand, was searched for graves in GPR reflection profiles. Thicker sandy areas of beach and dune sand are easily visible in profiles as low-amplitude reflection layers, sometimes with visible cross-bedding, and often exhibiting jumbled reflections because of the periodic activity of maintenance workers.

In areas surveyed close to the modern beach the gently dipping reflection surfaces from beach units are visible in profiles and the usual planar surface of the bedrock is broken into individual blocks, presumably created by storm erosion (**Figure 6.6**). Each of the bedrock cobbles and boulders that were eroded out of the intact bedrock produces a reflection hyperbola, and large areas of eroded bedrock can be readily identified in all the seaward locations of the study area.

In one of the large grids of GPR profiles, the edge of the high-amplitude oolite bedrock could be defined in amplitude maps. These high bedrock areas were then confirmed by interpreting individual reflection profiles. The amplitude slice-maps also show an area that contains only low amplitude planar reflections with many point-source reflection hyperbolas in the middle of the grid, surrounded by the bedrock. This is a small basin filled with beach and aeolian sand (**Figure 6.7**). In this small sandy basin, which is about 10 × 15 m in dimension, many graves are visible in profiles and on the amplitude slice-maps, each identifiable by individual reflection hyperbolas in profiles or high-amplitude features in the maps. The deeper slice from 120 to 160 cm shows the pipe that was placed in this area during the construction of the military barracks, when people commented on the exposure of human bones.

In this case, the deeper sandy areas could be easily differentiated from the areas where the oolite bedrock was close to the surface. These sand units were well laminated and show the distinct seaward sloping reflections common to beach sand, which inland are homogeneous and non-reflective suggesting well-winnowed aeolian sand. The location

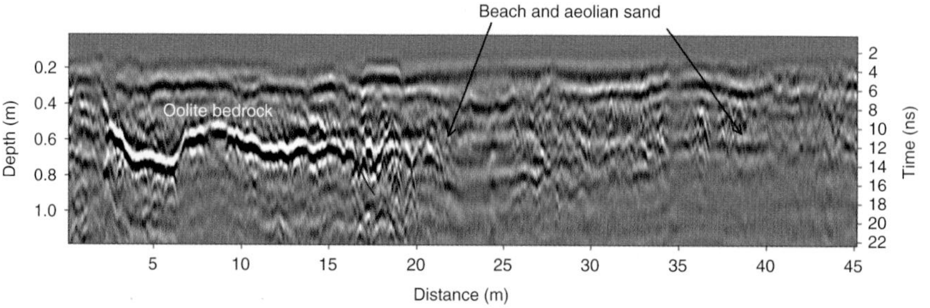

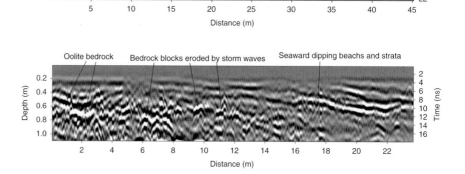

Figure 6.5 Eroded oolite bedrock along a beach in Key West, Florida, showing the distinct bedrock reflection in this 400 MHz reflection profile, with lower amplitude reflections generated in the beach and aeolian sand.

Figure 6.6 Beach units displayed in a 400 MHz reflection profile dip seaward from the eroded bedrock blocks in Key West, Florida.

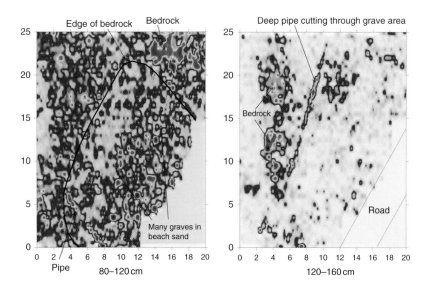

Figure 6.7 Amplitude slice-maps from Key West, Florida, showing the edge of eroded bedrock and many graves in a small basin covered by aeolian sand. From Conyers (2012). © Left Coast Press, Inc.

of this sandy basin surrounded by shallow bedrock corresponds to the area in oral histories of the burial ground, and makes a great deal of sense with respect to the amount of energy people would have expended in burying humans with little status or wealth.

Lagoon and Tidal Flats

Along sea cliffs in western Portugal aeolian dunes overlay a thin sequence of lagoon and beach sand, which today outcrops many tens of meters above sea level (**Figure 6.8**). These units are dated to about 36,250 BP during the late Pleistocene (Benedetti et al. 2009), and the units are of interest as Neanderthal tools have been found associated with this horizon (Bicho and Haws 2012; Haws et al. 2010). The thin near-shore sequence consists of a pebble-sand unit sitting on tidal flat mudstone. This sequence is well-cemented and forms a small resistant layer in the sea cliff outcrops. It is part of a regressive sequence that is then capped by a thin beach sand unit. Later, the area became further exposed and was covered by aeolian dunes. It was hoped that GPR profiles could define the tidal flat unit and associated beach areas in three dimensions and perhaps locate tidal channels or the location of shorelines where Neanderthals would have conducted activities and discarded artifacts. Before this could be done, it was important to see if this layer could be defined with GPR where energy had to travel through 10–20 m of aeolian dunes.

The first test of the GPR method to study these near-shore layers was conducted at a small outcrop where a sand dune covers the top of the beach cliff, near where the picture in **Figure 6.8** was taken. The GPR reflection profile (**Figure 6.9**) was corrected for topography and the 400 MHz reflections processed to display the reflections from the PV-3 unit (the local name for the tidal flat-beach unit). This profile was tied directly to the outcrop (using the method illustrated in **Figure 3.1**), so there was no doubt that the high-amplitude planar reflection was generated from the interface of these units and the overlying aeolian dunes. In this test, there was less than 1 m of sand overlying this interface, so the 400 MHz profile was capable of displaying this unit below the sand dunes and also shows areas of concentrated pebbles that produced many small reflection hyperbolas.

Pleistocene beach overlain by aeolian sand

PV-3 pebble-sand
on tidal flat mudstone

Figure 6.8 Tidal flat mudstone overlain by a very thin pebble-sand, encased in aeolian dunes on the west coast of Portugal.

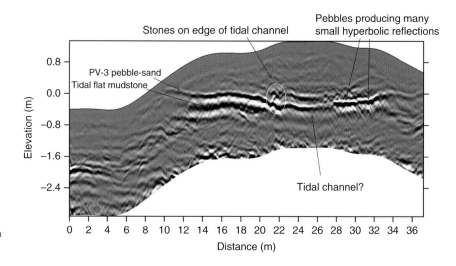

Figure 6.9 A topographically adjusted 400 MHz reflection profile displaying the tidal flat unit shown in **Figure 6.8** as a high-amplitude planar reflection surface.

Only the axes of these point-source hyperbolas generated from the pebbles were visible, as their apexes merged into one planar reflection (**Figure 6.9**).

A very shallow channel, which was likely produced on the tidal flat can be discerned, and some larger rocks that produced hyperbolic reflections along its edge (**Figure 6.9**). This test demonstrates how 400 MHz energy is more than capable of defining small features such as the pebbles in the tidal flat unit and the very shallow tidal channel on a thin near-shore unit of this sort. Unfortunately, this unit is covered everywhere else along this coastline by a very thick sequence of dunes reaching sometimes tens of meters. This test indicated that the 400 MHz energy could not penetrate more than about 3–4 m in this sand, making it unsuitable for mapping this unit along the rest of the coastline.

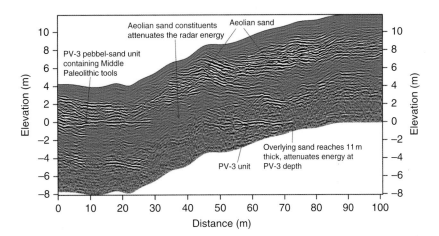

Figure 6.10 A topographically adjusted 270 MHz reflection profile showing the tidal flat unit pictured in **Figure 6.8** being progressively buried by aeolian sand away from the coastal cliffs.

To test how deeply buried thin tidal flat units of this sort could be potentially visible with GPR, a 270 MHz antenna was used to collect a profile over a dune reaching 20–25 m in thickness. The same PV-3 tidal flat unit was again correlated to reflections visible in the profile (**Figure 6.10**). The tidal flat horizon was still visible as a high-amplitude planar reflection as it was progressively covered by dune sand inland from the sea cliff. The 270 MHz profile did not provide the resolution that the 400 MHz profile could, but the surface was still visible until it was covered by 11 m of aeolian sand (**Figure 6.10**).

Along this transect the energy was attenuated in one area beneath some aeolian sand that attenuated propagating radar energy for some reason. Directly under some non-reflective dune sand, the PV-3 unit also disappears. What constituents might be found in this aeolian unit that produced the attenuation are not known, but it could be as simple as the addition of some clay that is a little more electrically conductive. When the PV-3 layer appears again in the profile further inland, the overlying sand units again show distinct reflection surfaces from the aeolian units. The 270 MHz antennas, while able to transmit and receive energy to 11 m depth in these dunes, did not produce reflected waves capable of defining aspects of the near-shore units necessary for this paleogeographic study in this very thick overburden sand.

Aeolian Dunes

Along the northern coast of Queensland, Australia, an area was studied with GPR that contains historic and likely prehistoric burials known to exist in coastal sand dunes (Sutton and Conyers 2013). The study area is directly inland from a salt water bay, covered with late Holocene sand dunes. These aeolian deposits rest directly on weathered bauxite bedrock. The area is well documented as a cemetery, which was used by Aboriginal people before a Presbyterian mission was built nearby in the last part of the 19th century. This burial ground continued to be used through the 20th century to bury both European and Aboriginal people using a variety of burial practices ranging from traditional bark cloth-wrapped bodies marked by coral stones to more elaborate European-style wooden caskets with headstones.

In preparation for fencing this cemetery area, marking the graves, and placing other monuments within this important cultural area, a number of GPR surveys

Figure 6.11 Low-altitude photo showing the coastal dunes, an interdune low area, and the beach in northern Queensland, Australia. The reflection profile A-A′ is shown in Figure 6.12.

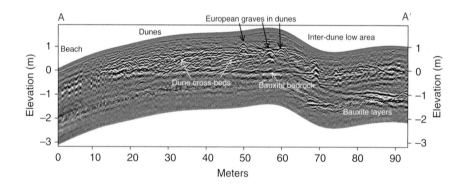

Figure 6.12 A topographically adjusted 400 MHz reflection profile that was collected from the beach onto coastal dunes and into an interdune area. A basal bauxite bedrock layer and overlying cross-bedded dunes are visible, which is the sediment matrix for graves.

were conducted. Low-altitude photos were taken from a tethered kite to produce an image of the ground on which the interpreted GPR reflection features could be placed (Figure 6.11). This photo readily shows the extent of the beach, coastal dunes, and an interdune low area that can often retain water during the summer rainy season. Inland from this photo is another band of dunes, which was not a part of this study.

A correlation profile was collected from the beach onto the dunes and into the interdune area (Figure 6.12). This reflection profile shows salt water attenuation on the beach, and a progressively thickening sand package of aeolian dunes just inland from the beach. These dunes show distinctive cross-bedding reflections resting directly on a highly reflective planar surface, which is the bauxite bedrock. The graves of "European style" burials, which are wooden caskets, are visible as reflection hyperbolas in the dunes. The interdune area also contains fine-grained carbonate sand, much like on the active dunes, but lacking cross-bedding reflections.

Amplitude slice-maps were constructed in a large grid of GPR data collected over the active dunes and into the interdune area (Figure 6.13). The slices were generated parallel to the ground surface, which was an excellent amplitude sampling method to

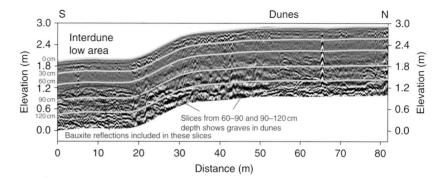

Figure 6.13 A topographically adjusted 400 MHz reflection profile showing the amplitude slices constructed using the profiles in the GPR grid shown in Figure 6.11.

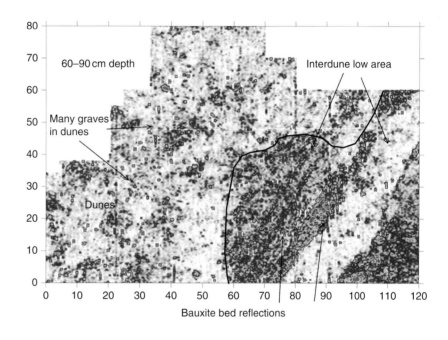

Figure 6.14 The 60–90 cm depth amplitude slice-map, the location of which is shown in Figure 6.11, displays the many graves in the dunes. When the slices cut through the interdune areas, the linear amplitude features display bedding variations in the bauxite bedrock.

produce maps of high amplitude graves in the active sand dunes (Figure 6.13). When the aeolian sand is thinner in the interdune areas these slices sampled and displayed the very high amplitudes of layers in the bauxite bedrock, generating many high-amplitude reflections.

In the 60–90 cm amplitude slice the tops of many graves that generated high-amplitude reflections are visible (Figure 6.14). Where that slice has crossed from the dunes into the bauxite bedrock, the linear high-amplitude reflections of that unit are highly visible and distinctly different than the dune reflections. This slice readily identifies the difference between the active dunes and the interdune area of thin sand. The area in the thick dune deposits display the highest amplitude reflections generated from the burials, and the cross-beds in the aeolian dunes were effectively invisible.

In the Jaguaruna region of the Santa Catarina State of southern Brazil many shell mounds contain artifacts and materials dating back about 9000 years before the present (Gaspar et al. 2008). The largest shell mounds are located near coastal lagoons, many on barrier islands and some inland, but all still in close proximity to the lagoons (Carvalho do Amaral et al. 2012). The lagoon system was more geographically extensive in the early Holocene, and has been progressively filling with sediment over time due to delta progradation from rivers flowing from the west and also sediment washed and blown in from beaches to the east. It has also been suggested that there was a sea level fall since the mid-Holocene, which partially drained the lagoon at the same time it was filling with sediment (Angulo et al. 2006). All these environment changes affected the productivity of the lagoon ecosystem and therefore human exploitation of animals and plants in and along its margin. This general area surrounding the back-barrier island lagoon contains a variety of depositional environments, with the most important for this study being the barrier island aeolian dune sequences adjacent to the lagoon.

Recent work using microfossils from cores suggest that the lagoon was progressively disconnected from the ocean with a reduction in the overall lagoon environment, beginning about 5000 BP (Carvalho do Amaral et al. 2012). Whatever the cause of these environmental changes, the reduction in biologically productive ecosystems must have affected people's ability to hunt, fish, and gather plant and animal resources. To better understand this interaction between people and the environment an integration of the geological record that defines environments of deposition with evidence of changing human adaptations can help explain these cultural changes as related to environmental changes along this coast over time.

More than a thousand shell mounds are known along the coast of Brazil, but unfortunately many of them have been destroyed by mining them for construction materials (Gaspar et al. 2008). These mounds contain a complex stratigraphy of layered shells, charcoal, possible occupation surfaces, human burials, hearths, and post holes from structures. Artifacts associated with the largest of these document sophisticated fishing and mollusk collection along the coastline and in the lagoons, with abundant grinding stones used for vegetable processing. While these mounds contain artifacts that document intensive marine resource exploitation, recent work suggests that the shells used in their construction were deposited there only in a secondary context and used as building material rather than primary waste from feasting or every day shellfish consumption. The mound construction occurred at a time when societies were becoming progressively more complex and the shell mounds were more likely monumental construction than solely disposal areas, as was originally suggested by archaeologists decades ago (Gaspar et al. 2008). They may have been built as raised sites for elite burials, places for ceremonial activities, or perhaps monuments displaying territory ownership by aspiring elites.

There is evidence of shell mound construction in the study area that started about 7500 years ago inland, with the largest of the shell mounds about 5000 years ago. Shell additions and modifications to these mounds occurred periodically over many centuries (Gaspar et al. 2008). Mound construction seems to have ceased about 2000 BP, with little evidence of intensive human activity throughout the area after that time. This decrease in human activity may be related to agricultural intensification inland that drew people away from the coast as new food products were available elsewhere. However, this cultural change could also be partially related to environmental changes along the coast, as the lagoon environment became less productive due to sediment infilling. The placement of people and their activities within these complex near-shore environments and an analysis of these changes over time are therefore crucial to a study of these interesting cultural changes over many thousands of years. The GPR method

Figure 6.15 Two of the largest Prehistoric shell mounds on the Brazil coast with the location of the GPR grids in the dunes in the foreground. Photograph by Tiago Attore. Reproduced with permission.

plays a role here because almost all of these habitation and activity sites are invisible below the dune sand. While the lagoon environment along this coast was still highly productive the locations where people lived and worked appear to have changed. A better definition of how people used a changing landscape is what stimulated this study with GPR.

The two largest and most intact shell mounds in Brazil are known as Figueirinha 1 and 2 (**Figure 6.15**). Figueirinha 1 is a typical monumental shell mound with abundant burials (Gaspar et al. 2008). Figueirinha 2, located nearby, is about 600 years younger and composed mostly of sand with fewer shell layers. Both show intensive human activity, but no occupation surfaces where everyday activities took place. No one had located the areas of habitation or other work areas nearby, so a number of GPR grids were collected in dunes adjacent to these large mounds to search for possible occupation surfaces or other cultural features below and within the aeolian deposits (**Figure 6.15**).

As a way to correlate sediment units and possible cultural horizons within the dunes GPR correlation profiles were collected where distinctive stratigraphic units were visible on the surface (**Figure 6.16**). One of these layers is a weathered surface that formed at the same level as lagoon water nearby, identifiable as an iron oxide impregnated weathering horizon (**Figure 6.17**). The antennas placed on this horizon and then moved over the adjacent modern dune and the weathered surface could be traced as it became buried progressively deeper.

This oxidized surface is almost perfectly horizontal and overlain by recent dune sand, and also an older dune dated at 4000 BP, which was stabilized by a shell pavement. This pavement surface is exposed elsewhere and interpreted as a shell-mantled living surface constructed by people who stabilized the dunes to produce

Figure 6.16 Iron oxide cemented layer that could be traced into the sand dunes as a way to correlate reflection horizons in the GPR profiles. Photograph by Tiago Attore. Reproduced with permission.

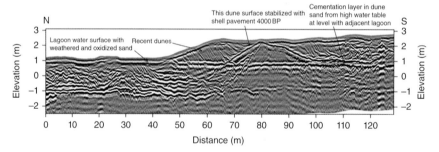

Figure 6.17 Topographically corrected 270 MHz reflection profile showing the iron-oxide weathering layer on which the antennas sat in **Figure 6.16**, covered by recent and older dunes dating to about 4000 BP. The weathering horizon is at the level of the adjacent lagoon, and is likely a ground-water related cemented layer that cross-cuts dunes of many different ages. Data from Tiago Attore.

work areas. There are also other dune units visible in the profile in **Figure 6.17**, but their age is unknown.

A general understanding of the age of many different aeolian dune packages was then compiled for the area, which are sometimes bounded by weathering surface and others by shell pavement layers that are often associated with charcoal, hearths, artifacts, and weak soil horizons, which formed with dune stabilization. These reflective surfaces are different enough from the aeolian sand to generate reflections as they retain more water and were used for correlation of dunes deposited during different time periods.

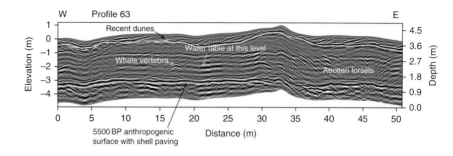

Figure 6.18 A 270 MHz reflection profile showing dated layers within aeolian quartz sand units. A whale vertebra produced a distinctive point-source reflection hyperbola. Data from Tiago Attore.

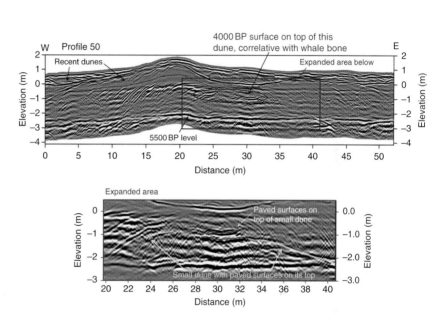

Figure 6.19 An intradune reflection surface was generated from shells that were used as a stabilization platform for work and living areas within these aeolian units. This layer is correlative to the horizon where the whale vertebra was uncovered, visible in **Figure 6.18**. Data from Tiago Attore.

A large grid of 270 MHz reflection profiles was collected in one of the dune areas where a correlation profile could trace the prehistoric stabilization surfaces, constructed from shells, and other reflective horizons of unknown origin. These profiles show excellent energy penetration of more than 6 m with good stratigraphic resolution in this quartz sand (**Figure 6.18**). This profile displays an aeolian unit about 3 m thick, bounded at the base by a 5500 BP shell-paved surface and at the top by recent dunes. On the eastern side of this profile the unit above the 5500 BP surface displays recognizable aeolian cross-bedding while much of the rest of this unit displays horizontal beds. Within this unit a striking point-source hyperbola is visible between 17 and 18 m (**Figure 6.18**), which is anomalous in this fine-grained sediment. It was excavated and found to be a whale vertebra resting on a weak soil zone or living surface that was not reflective in this location. The vertebra appears to have been placed there by people who were using these dunes, perhaps as an area for meat preparation. At about this same level the local water table was also encountered.

About 12 m south of the profile where the whale bone was visible (**Figure 6.18**), the aeolian dunes display the same general stratigraphy, but there are at least two distinct horizontal surfaces at the same level in the dunes as the whale vertebra to the north (**Figure 6.19**). This presumed living surface in Profile 50 (**Figure 6.19**) is bounded on

Figure 6.20 Photograph showing the shell paving surface producing the distinct reflection seen in **Figure 6.19**, which was a prehistoric method of stabilizing these sand dunes. Photograph by Tiago Attore. Reproduced with permission.

the east and west by sand units that are well cross-bedded, and may be part of later aeolian deposition that covered the site sometime after these surfaces were used. The horizontal living surfaces appear to have been constructed on a small dune about 4000 BP (**Figure 6.19**). In other adjacent profiles there are many other complicated reflections at this same level on what would have been the top of the small dune that was stabilized by human activity. This surface was excavated (**Figure 6.20**) and shows a weak soil development associated with large shells that were used as a paving surface. The shells are the detritus from food production, but they appear to have been placed on the dunes as a sand stabilization method.

Amplitude slice-maps constructed from the profiles in this grid show an abundance of reflections at all depths, most of which are displaying cross-beds in aeolian dunes (**Figure 6.21**). The portion of the sand dune package that contains the whale bone and the paved surfaces is displayed in the slice from 3.3 to 3.9 m. It is surrounded by non-reflective dunes that filled in this area after it was abandoned, which appear as white in this amplitude map. The objects on the surfaces and variations in the paving materials on the top of the dune are all displayed as high-amplitude reflections in the 3.3–3.9 m slice (**Figure 6.21**). These high amplitude reflections may have been generated from shells, charcoal, hearths, bones and other materials. There are many other likely work and living areas similar to the one excavated in the southwest portion of the GPR grid, which are unexcavated (**Figure 6.21**).

The GPR profiles were an excellent resource for mapping these coastal dunes that were modified by humans during the same time nearby monumental shell mounds were being built. It appears that people stabilized this aeolian landscape to provide working and living surfaces near the productive lagoon and beach areas. These can be identified with GPR within this complex sedimentary package and demonstrates one method people used to adapt to this complex and changing landscape. Using GPR previously unknown areas in this complex ecosystem that were used by these hunter-gatherer, marine resource-adapted people can be studied during a time of culture change.

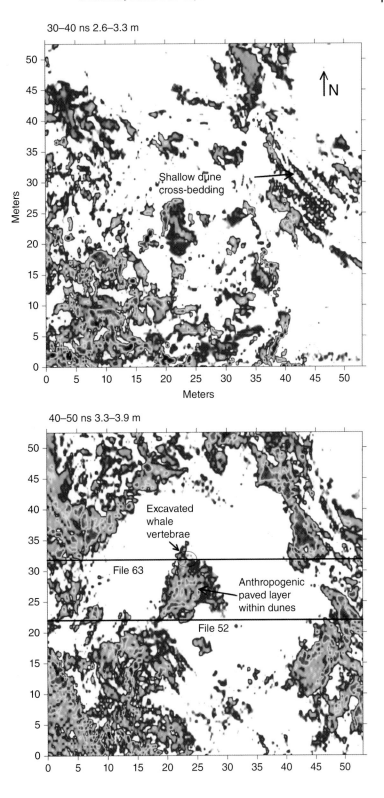

Figure 6.21 Amplitude slice-maps showing the living surface dated to about 4000 BP within the dunes in southern Brazil. The locations of Profiles 63 and 50 in Figures 6.17 and 6.18 are shown in the 40–50 ns slice. Data from Tiago Attore.

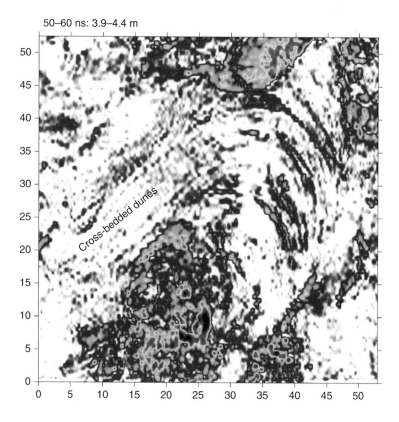

Figure 6.21 (Continued)

References

Angulo Rodolfo J., Lessa Guilherme C., & Souza Maria Cristina (2006) A critical review of mid- to late-Holocene sea-level fluctuations on eastern Brazilian coastline. *Quaternary Science Reviews*, vol. 25, pp. 486–506.

Ballard, Robert D., Stager, Lawrence E., Master, Daniel et al. (2002) Iron age shipwrecks in deep water off Ashkelon, Israel. *American Journal of Archaeology*, vol. 106, no. 2, pp. 151–68.

Benedetti, Michael M., Haws, Jonathan A., Funk, Caroline L. et al. (2009) Late Pleistocene raised beaches of coastal Estremadura, central Portugal. *Quaternary Science Reviews*, vol. 28, no. 27, pp. 3428–47.

Bicho, Nuno & Haws, Jonathan (2012) The Magdalenian in central and southern Portugal: human ecology at the end of the Pleistocene. *Quaternary International*, vol. 272, pp. 6–16.

Bristow, Charles S. & Pucillo, Kevin (2006) Quantifying rates of coastal progradation from sediment volume using GPR and OSL: the Holocene fill of Guichen Bay, south-east South Australia. *Sedimentology*, vol. 53, no. 4, pp. 769–88.

Carson, Mike T. (2011) Palaeohabitat of first settlement sites 1500–1000 BC in Guam, Mariana Islands, western Pacific. *Journal of Archaeological Science*, vol. 38, no. 9, pp. 2207–21.

Carvalho do Amaral, Paula Garcia, Fonseca Giannini, Paulo Cesar, Sylvestre, Florence, & Pessenda, Luiz Carlos (2012) Paleoenvironmental reconstruction of a Late Quaternary lagoon system in southern Brazil (Jaguaruna region, Santa Catarina state) based on multi-proxy analysis. *Journal of Quaternary Science*, vol. 27, no. 2, pp. 181–91.

Conyers, Lawrence B. (2012) *Interpreting Ground-penetrating Radar for Archaeology*. Left Coast Press, Walnut Creek, California.

Dickinson, William R. (2000) Hydro-isostatic and tectonic influences on emergent Holocene paleoshorelines in the Mariana Islands. Western Pacific Ocean. *Journal of Coastal Research*, vol. 16, no. 3, pp. 735–46.

Dickinson, William R. (2003) Impact of mid-Holocene hydro-isostatic highstand in regional sea level on habitability of islands in Pacific Oceania. *Journal of Coastal Research*, vol. 19, pp. 489–502.

Douglas, Bruce C. (1997) Global sea level rise. *Journal of Geophysical Research*, vol. 96, no. C4, pp. 6981–92.

Gaspar, Maria Dulce, DeBlasis, Paulo, Fish, Suzanne K., & Fish, Paul R. (2008) Sambaqui (shell mound) societies of coastal Brazil. In: Helaine Silverman & William Isbel (eds.) *Handbook of South American Archaeology*, pp. 319–35. Springer, New York.

Haws, Jonathan A., Benedetti, Michael M., Funk, Caroline L. et al. (2010) Coastal wetlands and the Neanderthal settlement of Portuguese Estremadura. *Geoarchaeology*, vol. 25, no. 6, pp. 709–44.

Master, Daniel M. (2003) Trade and politics: Ashkelon's balancing act in the seventh century BCE. *Bulletin of the American Schools of Oriental Research*, vol. 330, pp. 47–64.

Møller, Ingelise & Anthony, Dennis (2003) A GPR study of sedimentary structures within a transgressive coastal barrier along the Danish North Sea coast. In: Charles S. Bristow & Harry M. Jol (eds.) *Ground Penetrating Radar in Sediments*, pp. 55–65. Geological Society Special Publications 211, Geological Society, London.

Moore, L.J., Jol, Harry M., Krus, Sarah, et al. (2004) Annual layers revealed by GPR in the subsurface of a prograding coastal barrier, southwest Washington, USA. *Journal of Sedimentary Research*, vol. 74, no. 5, pp. 690–6.

Moshier, Stephen O., Master, Daniel, Lepori, Jacob, et al. (2011) Geological foundations of ancient seaport, Ashkelon, Israel. *Abstracts of the Geological Society of America Conference*, Oct. 9–12, 2011, Minneapolis, Minnesota.

Muhs, Daniel R., Simmons, Kathleen R., Schuman, Randall, & Halley, Robert B. (2011) Sea-level history of the past two interglacial periods: new evidence from U-series dating of reef corals from south Florida. *Quaternary Science Reviews*, vol. 30, no. 5, pp. 570–90.

Neal, Adrian & Roberts, Clive L. (2000) Applications of ground-penetrating radar (GPR) to sedimentological, geomorphological and geoarchaeological studies in coastal environments. *Geological Society*, vol. 175, no. 1, pp. 139–71.

Pearl, Frederic B. & Sauck, William A. (2014) Geophysical and geoarchaeological investigations at Aganoa Beach, American Samoa: an early archaeological site in Western Polynesia. *Geoarchaeology*, vol. 29, no. 6, pp. 462–76.

Peltier, W. Richard (2002) On eustatic sea level history: last Glacial Maximum to Holocene. *Quaternary Science Reviews*, vol. 21, no. 1, pp. 377–96.

Sutton, Mary-Jean & Conyers, Lawrence B. (2013) Understanding cultural history using ground-penetrating radar mapping of unmarked graves in the Mapoon Mission Cemetery, Western Cape York, Queensland, Australia. *International Journal of Historical Archaeology*, vol. 17, no. 4, pp. 782–805.

Waters, Michael R. (1992) *Principles of Geoarchaeology, A North American Perspective*. The University of Arizona Press, Tucson, Arizona.

7 Lakes and Deltas

Abstract: Freshwater lakes are an excellent medium for GPR as there is almost no energy attenuation in the water column and high-amplitude reflections are generated from the lake bottom and sediment layers below. Data collection must be from a boat with locations taken by GPS. Important stratigraphy and buried features can be seen in subwater stratigraphy in these otherwise inaccessible environments. Deltas that were formed on lake margins are visible in reflection profiles and amplitude maps as prograding forset beds or coarser grained sediment that contrasts with finer grained units in the deeper lacustrine environment.

Keywords: lakes, deltas, freshwater, landslides, boat collection

Introduction

Fresh water lakes are an excellent medium for GPR studies, as water with low concentrations of dissolved salt is electrically resistive and readily allows the passage of radar waves with negligible attenuation (Conyers 2012, p. 74). Lakes are any closed body of water that has a natural dam that has allowed water to accumulate for any length of time. They are filled with clastic, chemical, and organic sediments and, if allowed to fill for a long time in the right conditions, can be transformed into swamps and bogs (Chapter 5). Deltas accumulate on the margins of these bodies of water wherever sediment is being transported into the basin. Both lakes and the deltas that form on their margins will be covered in this chapter.

Lakes have attracted humans for millinea, as they are areas of rich faunal and floral resources for hunters and gatherers and fishers, with locations along their margins providing abundant grazing land and fertile soils for agriculture (Ismail-Meyer et al. 2013; Ruoff 2004). Habitation structures were often built on their margins, and even on stilts and platforms in shallow water near the shore. The lake sediments are an excellent medium for the preservation of organic matter discarded by people living in these dwellings as the lake floor is an anoxic environment that precludes deterioration.

Geoarchaeology including GPR analysis can help understand lake level changes over time, especially if there has been erosion and depositional changes due to climatic fluctuations when lake bottoms periodically dry out and then refill again. An analysis of differing environments along lake shorelines can also be accomplished by studying

Ground-penetrating Radar for Geoarchaeology, First Edition. Lawrence B. Conyers.
© 2016 John Wiley & Sons, Ltd. Published 2016 by John Wiley & Sons, Ltd.

depositional units and placing people within those ancient landscapes. While lakes are not as subject to dynamic geological processes as fluvial and coastal environments are, they still exhibit changes that can be studied geologically and with the application of GPR (Digerfeldt et al. 2007; Magny 2004).

The use of GPR has been somewhat limited in lakes perhaps because data must be collected from watercraft and positioning of reflections transects in space can be problematic. Some notable successes have been published from the Swiss lakes where excellent energy penetration was noted in lake bottom sediments, with positioning accomplished using GPS (Fuchs et al. 2004). Collecting GPR data on frozen lakes can also be an excellent way to overcome positioning obstacles (Conyers 2012, p. 74), and excellent radar penetration is obtained from ice and snow covering on lakes (Arcone et al. 2006). Glacial lakes are often best for GPR as the water is usually very fresh with a few dissolved solids that might cause water to attenuate energy, or in the lake bottom sediment (Johnsen and Brennand 2004; Sambuelli and Bava 2012).

Lakes

At an elevation of more than 10,000 feet (3000 m) in western Colorado a number of lakes have been active depositional settings since the late Pleistocene (Cole and Sexton 1981). Some of them appear to be dammed by glacial moraines, and others by landslide deposits from the surrounding high mountains (**Figure 7.1**). One lake on Grand Mesa is bounded by angular boulders that were deposited on steep talus slopes along one margin, with

270 MHz antennas
in canoe bottom

Landslide rubble

Collecting GPR profiles from canoe

Figure 7.1 The 270 MHz antennae placed in the bottom of a canoe were used to collect reflection profiles as the boat was moved along a rope floating on the lake surface to assure linear transects. This lake at high elevation in the Rocky Mountains of Colorado is dammed by glacial moraine and bounded on one side by landslide rubble. Conyers (2012). Reproduced with permission of Left Coast Press, Inc..

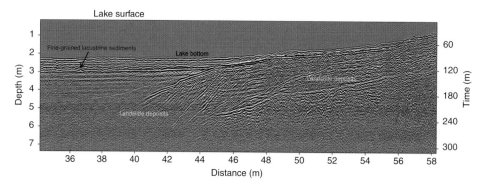

Figure 7.2 A 270 MHz reflection profile collected on the lake shown in **Figure 7.1**, displaying landslide events on the lake margin and fine-grained sedimentation toward the middle of the lake.

glacial moraine till forming the dam on the opposite bank. In preparation for coring the lake sediments for paleoenvironmental analyses GPR data were collected using 270 and 400 MHz antennae placed in the bottom of a fiberglass canoe, with **survey transects** measured by ropes floating in the water. The beginning and end of profiles was located in space by GPS. Reflection traces were collected in time rather than in distance and **fiducial marks** were input in the data string every 10 m along the ropes in order to correct for variations in the canoe speed across the lake. Later during data processing all traces were put into corrected space using the 10 m marks. The same type of collection could be done by placing traces into space using GPS location data that were collected at the same time as the GPR reflection traces (Conyers 2013, p. 31).

Excellent radar wave penetration was visible on all the profiles collected in this lake, with 3–4 m thickness of high-amplitude reflections displayed in the lake bottom sediments (**Figure 7.2**). On the lake margin closest to the steep slope containing landslide rubble (**Figure 7.1**) there are a number of wedges of these thick units consisting of many point-source reflection profiles generated from the angular boulders. Each distinct landslide event is bounded by laterally extensive planar reflections, which are likely deposits of finer grained clastic sediments deposited during normal lake sedimentation. About 20–30 m away from the shoreline, these thick rubble deposits are replaced by fine-grained lacustrine beds in horizontal well-bedded sequences typical of most lakes. Individual reflections from boulders in the fine-grained horizontally bedded units are probably from large stones that were "rafted" out into the middle of the lake by floating ice, and dropped to the lake bottom when the ice melted.

Deltas

A shallow lake in the Dominican Republic was tested with GPR as a way to map the changing delta location and also variations in other near-shore depositional environments that had been noticed in historic aerial photos. This lake, while composed of fresh water, is quite cloudy and appears to have a good deal of sediment and other dissolved solids in the water. Near the border with Haiti, on the north coast of the island of Hispanola (**Figure 7.3**), this lake appears to be dammed by some type of neotectonic event, which created a fault-controlled basin. Water flows into the basin today from small deltas on

Mountains on the border with Haiti

400 MHz antennas and control system placed in bottom of fiberglass boat

Figure 7.3 The 400 MHz antennae were placed in the bottom of this boat and transects were collected at a set rate across the surface of the lake.

the northeast and southwest. A sequence of historic aerial photos indicates that these deltas have changed position over time during flooding episodes in intense summer and autumn storms.

Reflection profiles were collected by placing 400 MHz antennae in the bottom of a motorized boat that moved at a constant velocity across the lake. The locations of the beginning and end of profiles were collected with a handheld GPS. All reflection traces were then spatially corrected to place all reflection profiles into a UTM coordinate system. Some of the profiles were as long as 600 m. No topographic corrections were necessary and each of the reflection profiles displays the water–sediment interface as a high-amplitude reflection. The now invisible and overgrown delta from the 1960s on the west bank of this lake is visible in the reflection profiles as a 1 m thick sequence of inter-bedded coarse and finer grained sediments (**Figure 7.4**). Toward the deeper portion of the lake the delta sediments transition to a thin high-amplitude reflection near the lake bottom. These are finer grained lacustrine sediments deposited away from the coarser materials deposited in the delta.

Once the profiles were placed in space using the GPS coordinates, the amplitudes of the reflections in all the lines could be mapped. Discrete *x* and *y* coordinates were assigned to all amplitudes in all profiles and an average amplitude for all reflections recorded from the lake bottom to 20 ns within the sediment package was sampled and plotted. This calculation was made over a slice of sediment on the lake bottom equal to about 1 m in thickness. The absolute value of the average of the radar wave reflection strengths from the numerous interfaces between coarse and fine-grained sediment in the delta produced an overall high-amplitude value (**Figure 7.5**). The much thinner fine-grained

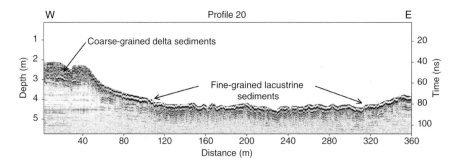

Figure 7.4 Reflection profile across the lake in Dominican Republic displaying delta deposits on the west and deeper lake sediments on the right. The location of this profile 20 in the lake is shown in **Figure 8.5**. Conyers (2012). Reproduced with permission of Left Coast Press, Inc.

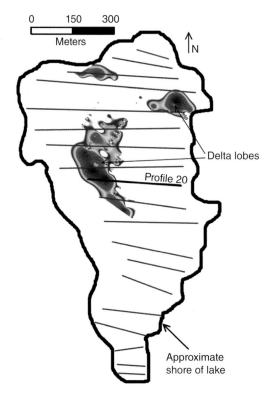

Figure 7.5 Reflection amplitude map of the sediments in the upper meter of lake bottom units, showing only the highest values obtained from alternating coarse and fine-grained sediments in two delta lobes on the west and northeast. One small high-amplitude area on the north was sampled from only one profile and its origin is unknown. From Harry M. Jol.

sediment in the deeper lake sediments produced only one high-amplitude reflection at the lake bottom, and below that level an average reflection amplitude for the sediment was much lower than the areas where the deltas had deposited alternating layers. The map in **Figure 7.5** therefore shows an overall sediment package amplitude with the high areas

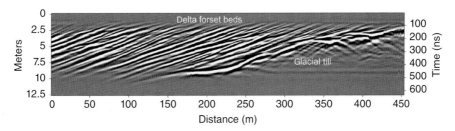

Figure 7.6 Reflection profile collected with 12.5 MHz antennae in a glacial lake in northern British Columbia displaying delta forset beds of a prograding sedimentary sequence lapping on to glacial till. Data from Harry M. Jol.

being the locations of thicker alternating coarser and finer sediments. The large lobe of the delta on the west portion of the lake does not continue all the way to the shore in the map in **Figure 8.5** because the boat's keel was rubbing on the shallow lake bottom there, and the GPR data were extremely noisy and had to be removed from the amplitude map.

Thicker delta deposits in lakes, especially where stream flow brings down pulses of coarse sediment periodically can produce large delta forset beds (Heteren et al. 1998), visible in GPR profiles (**Figure 7.6**). In a glacial lake in northern British Columbia each of these steeply dipping reflections show a progradation of sediment on top of glacial till (Buynevich et al. 2008). This profile was collected with very low-frequency 12.5 MHz antennae, producing good resolution to about 10 m depth in a very freshwater lake.

References

Arcone, Steven A., Finnegan, David C., & Liu, Lanbo (2006) Target interaction with stratigraphy beneath shallow, frozen lakes: quarter-wave resonances within GPR profiles. *Geophysics*, vol. 71, no. 6, pp. K119–31.

Buynevich, Ilya V., Jol, Harry M., & FitzGerald, Duncan M. (2008) Coastal environments. In: Harry M. Jol (ed.) *Ground Penetrating Radar Theory and Applications*, pp. 299–322. Elsevier Science Limited, Kidlington.

Cole, Rex D. & Sexton, John L. (1981) Pleistocene surficial deposits of the Grand Mesa area, Colorado. In: R. C. Epis & J. F. Callender (eds.) *Western Slope Colorado: New Mexico Geological Society Guidebook, 32nd Field Conference*, pp. 121–6. Albuquerque, New Mexico.

Conyers, Lawrence B. (2012) *Interpreting Ground-penetrating Radar for Archaeology*. Left Coast Press, Walnut Creek, California.

Conyers, Lawrence B. (2013) *Ground-penetrating Radar for Archaeology*, 3rd Edition, Altamira Press, Rowman and Littlefield Publishers, Lantham, Maryland.

Digerfeldt, Gunnar, Sandgren, Per, & Olson, S. (2007) Reconstruction of Holocene lake-level changes in Lake Xinias, central Greece. *The Holocene*, vol. 17, no. 3, pp. 361–7.

Fuchs, Michaël, Beres, Milan, & Anselmetti, Flavio S. (2004) Sedimentological studies of western Swiss lakes with high-resolution reflection seismic and amphibious GPR profiling. In: Evert Slob, Alex Yaravoy, & Jan Rhebergen (eds.) *Ground Penetrating Radar, 2004, Proceedings of the Tenth International Conference on GPR, Delft, Netherlands*. pp. 577–80. IEEE, Piscataway, New Jersey.

Heteren, Sytze Van, Fitzgerald, Duncan M., Mckinlay, Paul A., & Buynevich, Ilya V. (1998) Radar facies of paraglacial barrier systems: coastal New England, USA. *Sedimentology*, vol. 45, no. 1, pp. 181–200.

Ismail-Meyer, Kristin, Rentzel, Philippe, & Wiemann, Philipp (2013) Neolithic lakeshore settlements in Switzerland: new insights on site formation processes from micromorphology. *Geoarchaeology*, vol. 28, no. 4, pp. 317–39.

Johnsen, Timothy F. & Brennand, Tracy A. (2004) Late-glacial lakes in the Thompson Basin, British Columbia: paleogeography and evolution. *Canadian Journal of Earth Sciences*, vol. 41, no. 11, pp. 1367–83.

Magny, Michel. (2004) Holocene climate variability as reflected by mid-European lake-level fluctuations and its probable impact on prehistoric human settlements. *Quaternary International*, vol. 113, no. 1, pp. 65–79.

Ruoff, U. (2004) Lake-dwelling studies in Switzerland since Mellen 1854. In Francesco Menotti (ed.) *Living on the Lake in Prehistoric Europe*, pp. 9–21. Routledge, London and New York.

Sambuelli, Luigi & Bava, Silvia (2012) Case study: a GPR survey on a morainic lake in northern Italy for bathymetry, water volume and sediment characterization. *Journal of Applied Geophysics*, vol. 81, pp. 48–56.

8 Caves and Rock Shelters

Abstract: Sedimentary sequences in caves and rock shelters often hold a long record of important cultural materials, which because of the protection the enclosures afford, offer uniquely well-preserved artifacts. Layered floor units can be very complex stratigraphically as they often are highly disturbed by human modification actions but also biological and geological events. Hearths, anthropogenic fill sequences, discard areas and constructed floors are just a few of the layers and features visible in GPR profiles and amplitude maps. Stratigraphic units just outside these natural shelters can also be studied and anthropogenic modification of the surrounding landscape can be analyzed with GPR.

Keywords: caves, rock shelters, floors, hearths, apron, back-fill

Introduction

Caves and rock shelters provide benefits for archaeological research because they are the location of sediments and associated artifacts and anthropogenic features, which are not as readily eroded, weathered, or disrupted as open air sites. They can hold an archive of materials not otherwise discoverable in other contexts (Woodward and Goldberg 2001), sometimes containing rarely preserved perishable organic materials (Goebel et al. 2003; Murphy and Mandel 2012). Important floor sediments within caves and shelters can be sealed with roof-fall stones, and buried and preserved by the deposition of other sedimentary units. The resulting record can be a thick and culturally rich geological sequence holding evidence of multiple occupations. Careful analysis of the preserved geological strata in these protected, but complex environments, can often be defined with careful GPR analysis. When specific stratigraphic units and the artifacts contained in them can be placed within a temporal context the resulting geoarchaeological record can help in the understanding of cultural changes and many human behaviors from a record not available elsewhere.

The disadvantages common with many caves and rock shelters is that they were the location of intensive use, often for long periods of time (Goldberg and Macphail 2008). Human modification of floor surfaces and the removal and addition of materials often complicate stratigraphy. Some cave and rock shelter deposits can reach thicknesses of many tens of meters with deposition spanning millennia. These very thick sedimentary packages, which often consist of thinly layered strata, can be extraordinarily difficult to study using common GPR methods.

Ground-penetrating Radar for Geoarchaeology, First Edition. Lawrence B. Conyers.
© 2016 John Wiley & Sons, Ltd. Published 2016 by John Wiley & Sons, Ltd.

This is especially the case at depths greater than 3–4 m where low-frequency antennae must be used for depth penetration, with a resulting decrease in stratigraphic resolution. The ability to define important units within the upper few meters of thinly layered floor strata can still be excellent using GPR antennae in the 400 MHz range. The GPR method therefore provides an excellent tool for understanding many of these important buried deposits.

Rock Shelters

In the eastern plains of Colorado a small rock shelter provided a well-protected south-facing location for people from the late Archaic period (3000 BP) through much of the Plains-Woodland period (until about AD 850). The shelter holds the record of people who were in the process of transitioning from hunting and gathering to experimentation with agriculture, and toward the end of this long time span their lifestyles tended to become more sedentary (Gilmore 2008). This site is named Cherry Creek Canyon rock shelter for the small perennial drainage that flows about 20 m south of the shelter. One large trench was excavated in 1956, but the results of those excavations were not published and are available only in handwritten form, archived in the University of Denver Museum of Anthropology. Additional excavations were conducted in 2000 (Tchakirides 2002) that integrated GPR with these early excavation data confirming the presence of features and strata using GPR and additional test excavations. More GPR reflection data were acquired in 2014 and all previous datasets were integrated and reinterpreted.

The total area of the shelter within the drip line, and therefore protected from the weather, is about 300 m². This small area could have at most been a habitation for a family or perhaps a small group of 10–15 people. Both 400 and 900 MHz antennae were used to acquire reflection profiles inside the shelter and on a possible "discard apron" in front of the shelter that slopes downward toward the floodplain of Cherry Creek (Figure 8.1). The ground surface within the shelter is mostly flat, with a few minor undulations from recent activities including campfires set by recent hunters and perhaps the remains of the poorly back-filled 1956 excavations, where some excavated sediment remains in a few small piles. All profiles were placed into space and corrected for topography with a total station survey.

While the 900 MHz reflection profiles provided good stratigraphic resolution, depth penetration of that frequency radar energy in the floor sediment was only about 50 cm or so, which was not deep enough to resolve some of the important layers known to exist. Instead, the 400 MHz profiles provided the most useful dataset for geoarchaeological analysis. All reflection profiles were first resampled and gridded to produce 20 cm thick amplitude maps (Figure 8.2). These slices display a wealth of very complex reflections that are difficult to interpret from these images alone so that each reflection profile needed to be interpreted individually. The reflections in the profiles were then correlated to the stratigraphy known from excavations and visible reflection features were tied to those visible in the amplitude maps.

A number of reflection profiles displayed sloping planar reflections that dip toward the floodplain of the creek (Figure 8.3). Some of these are inside the drip line, which suggests that humans had renovated the floor surface, filling low areas to produce a flat floor surface totally within the rock overhang. The sloping surface visible in File 48 (Figure 8.3) is likely generated by reflection from a buried soil horizon, or from layers of differing materials dumped in the front of the shelter to level the floor. This sloping high-amplitude planar surface in front of the shelter has not been tested by excavations, so its origin is speculative.

Figure 8.1 Collecting 400 MHz reflection profiles in the Cherry Creek Canyon Rock Shelter of Colorado.

Inside the drip line a number of hearths are visible as high-amplitude reflections that could be correlated directly to superimposed hearths discovered in excavations (**Figure 8.4**). Near the back wall a storage cistern or other pit of some sort is visible cutting through well-layered units of floor fill. The age of this feature is not known, but its stratigraphic position indicates that it was incised late in the occupational history of this shelter. A high-amplitude planar reflection surface is visible at about 50 cm depth, which is known from excavations to be a charcoal-rich horizon that is about 10–15 cm thick, which contains many artifacts (Tchakirides 2002). Associated with this charcoal horizon are a number of shallow hearths known from excavations, which cannot be discriminated from the charcoal horizon reflection in GPR profiles. The artifacts found in the charcoal horizon are from the Plains-Woodland period dating between about 1700 and 800 BP.

This charcoal-rich layer unit was exposed in the 1956 trench and also in 1×1 m excavations from the 2000 field work (**Figure 8.2**). It is an important horizon as it is suggestive of long-term habitation and intensive human activities involving much wood burning. The artifacts within this unit are mostly stone and bone tools for hide working, hunting, and other domestic activities. Underlying the charcoal sediment layer is a thin clay layer, which appears to have been a deliberate import of clay into the shelter to produce a compact floor surface. This likely happened during a renovation episode when people modified a large floor surface for long-term habitation (Tchakirides 2002). The interface between the clay and the charcoal horizon produces the high-amplitude planar reflection visible in many 400 MHz profiles (**Figure 8.3**).

Just 1.5 m east of Profile 48 a reflection profile clearly shows a high-amplitude planar reflection from the southward sloping bedrock surface near the back of the shelter (**Figure 8.4**). The charcoal living surface can be seen intersecting this bedrock surface, with a number of small rocks sitting on the floor horizon at the very back of the shelter. The 400 MHz profile displays many weak point-source hyperbolas elsewhere on the floor horizon. These reflections were generated from individual cobble-sized or larger

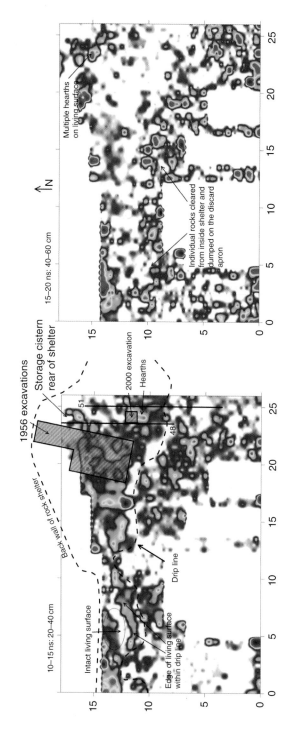

Figure 8.2 Amplitude slice-maps of 400 MHz data at the Cherry Creek Canyon Rock Shelter, Colorado. The excavations and location of profiles in Figures 8.3 and 8.4 are shown, with locations of the excavations, the drip line, and the rear wall of the shelter.

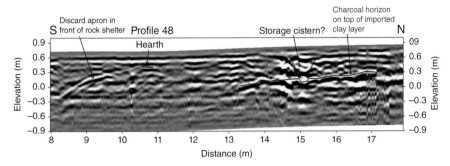

Figure 8.3 A 400 MHz reflection profile collected at the Cherry Creek Canyon Rock Shelter displaying high-amplitude planar reflection from the charcoal horizon within the shelter, and a possible discard apron or leveling fill deposit toward the floodplain of the creek to the south. Location of this profile is shown in **Figure 8.2**.

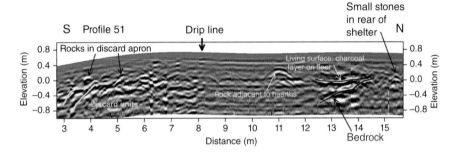

Figure 8.4 A 400 MHz reflection profile showing a number of sediment discard units toward the floodplain to the south, and the bedrock and charcoal living surface within the shelter.

stones that may have been used as seats, work areas, tools, or associated with hearths (**Figure 8.4**). Many stones of this sort were excavated from the charcoal-rich horizon and also from deeper levels in this shelter, beyond the depth of the 400 MHz energy penetration.

Many discard units sloping into the floodplain to the south of the shelter are also visible in GPR reflection profiles (**Figure 8.4**), which in this location are different from those visible in Profile 48 (**Figure 8.3**). Here, the discard apron is far outside the drip line, and these are probably layered trash deposits as opposed to the sloping surfaces in Profile 48 within the shelter overhang, which were likely produced by leveling operations to expand the floor surface.

Once the basic stratigraphy of the rock shelter was defined by correlating radar reflections to the sediment units known from the excavations, the complex reflections visible in the amplitude slice-maps could then be interpreted (**Figure 8.2**). In the western portion of the shelter a line of rocks just inside the drip line can be seen in the profiles, visible in the 20–40 cm amplitude map as a high-amplitude arc-shaped feature. This is the foundation of a wall consisting of stones that formed a barrier between a constructed floor surface within the shelter and the area outside the drip line. Inside this barrier an area of no-reflection about 2 m wide is visible in the slice-map from 20 to 40 cm depth between the drip line and the rear of the shelter. This small enclosed portion of the overhang is separated from the larger floor space to the east where a more extensive floor and hearths are visible. This small enclosure was likely used as a covered storage area.

Down the slope in the western portion of the surveyed area many rocks that are cobble sized or larger are visible in the 40–60 cm slice, displayed in an arc-shaped apron of high-amplitude point-source reflections (**Figure 8.2**). These were probably stones from ceiling fall or rocks brought into the shelter by humans that were later cleared out of the shelter during renovations. The stone clearance probably occurred during the Plains-Woodland occupation, coincident with or just before the shelter was renovated for longer term habitation. This is when the clay floor was constructed and the possible walled storage location on the west side was built. There were many fires lit inside the shelter soon after this, with the deposition of the charcoal horizon over the imported clay floor.

Artifact types found within this charcoal-rich floor sediment demonstrate that the inhabitants of the shelter can be placed within the Plains-Woodland period. This was a period when the hunting and gathering groups in this area of Colorado started to become less mobile coincident with experimenting with maize agriculture (Gilmore 2008). One small group of these people likely took "possession" of this shelter, investing a good deal of time to renovate it for longer term habitation. The charcoal layer on the clay floor surface indicates many fires over a long period of time, perhaps from year-round or winter habitation. Also, at this time people began to produce pottery, which also demonstrates a changing adaptation of these people to this area of Colorado, with settled partially agricultural people living in small shelters of this sort.

The excavations conducted at Cherry Creek Canyon Rock Shelter in 1956 penetrated below the charcoal living surface and found artifacts of late Archaic age, but in dramatically lower numbers. These deep horizons are unfortunately not visible in the GPR profile and were very poorly defined stratigraphically in the excavations, so these units would be poor candidates for GPR mapping. The early occupation of the shelter was probably by people who used this shelter only periodically during hunting and gathering expeditions, and had little motivation to modify and renovate this natural landscape feature. Only when people had agricultural fields nearby during the Plains-Woodland period were they motivated to modify this site with floor and wall construction, and it is these features that are visible with GPR.

Mapping Adjacent to Rock Shelters

In the outback of Queensland, Australia, a small south-facing rock shelter (**Figure 8.5**) was excavated in 2006 and 2008, which yielded age dates as old as 28,000 BP in a well-stratified sequence of sediments more than 1.8 m thick (Wallis et al. 2009). Three levels of artifact-rich sediments were identified, including expedient stone tools, pigment fragments, and many ceiling-fall stones. This shelter has very little overhang and likely formed more of a shade shelter than an overhang shielding occupants from the rain. A GPR survey was conducted to the south of the excavations in order to determine if there are possible cultural horizons in a small bedrock basin in front of the site, which was hypothesized to have been an area where sediment and artifacts would have collected over a long period of time.

The 400 MHz reflection profiles exhibit excellent energy penetration up to 3 m deep, with a distinct planar bedrock reflection (**Figure 8.6**). The bedrock reflection was visible during data collection and tested in shallow parts of the grid with an iron probe to calibrate for depth. Very prominent buried bedrock blocks can be seen in the profile with well-defined upper surfaces. These large blocks are much like those visible

Figure 8.5 Excavations at the Gledswood Rock Shelter, Queensland, Australia. The GPR grid was collected in front and to the right of the excavations. Conyers (2012). Photograph by Lynley Wallis. Reproduced with permission of Left Coast Press, Inc.

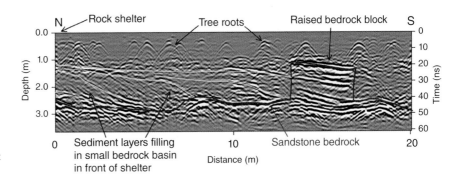

Figure 8.6 Bedrock layers and large bedrock blocks are visible as high-amplitude reflections, with prograding sediment layers that filled a basin in front of the shelter over time.

in much of the surrounding landscape (Conyers 2012, p. 66). Many sediment layers can be seen "prograding" out from the rock shelter, which filled a bedrock basin over time. Whether these are culturally rich layers or just geological units cannot be determined from the GPR profiles. However, they project directly into the excavations of the shelter and are generally correlative to the artifact-rich units uncovered there (Wallis et al. 2009).

Elevations of the bedrock reflection were then collected in all profiles within the grid to prepare a bedrock topography map in the basin in front of the shelter (**Figure 8.7**). This map shows this basin to be quite complex and not a "bowl"-shaped sediment trap as was expected. Instead, it is a series of small low areas, interrupted by large buried blocks of bedrock, such as the one visible in the reflection profile in **Figure 8.6**. During the early occupations, especially in the late Pleistocene (28,000 years ago), the area in front of the shelter was composed of flat-topped sandstone blocks, with intervening low areas. These low areas surrounded by higher bedrock exposures likely provide sediment traps, while the horizontal bedrock surfaces may have provided work areas for the people who periodically used this area.

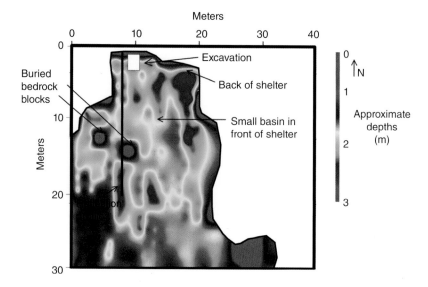

Figure 8.7 Depth to bedrock map in front of the Gledswood Rock Shelter, Queensland, Australia.

Caves

In the coastal mountains of Portugal a deep cave has been studied periodically since the 1950s, with intensive almost yearly excavations conducted starting in 1994 (Bicho et al. 2006; Haws 2012). Called Lapa do Picareiro, this small cave is about $10 \times 12\,m$ in dimension and had been excavated to about 10 m depth in 2014. At that depth cultural materials are still being discovered with every new excavation season. Age dates obtained from 7.4 m depth are from 45,700 BP, which are middle Paleolithic and appear to have been deposited by Neanderthals. Dates at the top of this cave floor sequence are from just the last few centuries, so this cave provides a very thick stratigraphic sequence that spans a very long time.

The sediment in the cave is primarily limestone breccia, with large blocks of ceiling fall interbedded with a few travertine units (Bicho et al. 2006). A number of 270 MHz profiles were collected on small steps within the cave in the hope that the bottom of the cave sediments could be found with GPR, and also to test the depth of energy penetration in this material (**Figure 8.8**). Radar energy was capable of being transmitted to about 8 m before it was attenuated (**Figure 8.9**). The base of the sediment package was not visible and no bedrock layer was visible. Energy began to be attenuated at about 6 m depth, with only some vague reflections recorded from the last 2 m in the sediment units. The GPR profile from within the cave shows a generally horizontally layered sequence of sediments, with no large features resolvable with this antenna.

The topography and buried landscape just outside the Lapa do Picareiro cave entrance is also interesting, which has been covered by a large volume of limestone breccia dumped there over decades of excavation. There is a vague remnant of a wall, perhaps part of a shepherd's structure, in front of the cave that is mostly covered with cave-excavated breccia. It was posed that there might be an apron of sediment that dates prior to excavations that is still present in front of the cave under this overburden. To map the pre-excavation surface of the ground two GPR reflection profiles were collected to see if the historic ground surface could be seen under the back-fill. The present topography of the small area

Figure 8.8 Collecting 270 MHz reflection data inside the Lapa do Picareiro cave in western Portugal to test the depth penetration in limestone breccia cave floor sediments.

270 MHz antennas

Six meters of cave fill exposed along this face

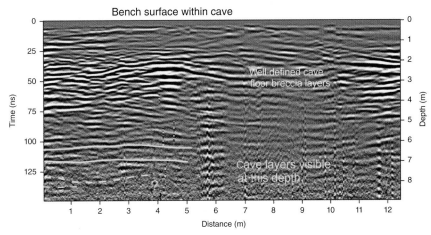

Bench surface within cave

Well-defined cave floor breccia layers

Cave layers visible at this depth

Figure 8.9 A 270 MHz reflection profile showing good energy penetration to about 6 m in the limestone breccia cave sediment at Lapa do Picareiro cave, Portugal.

in front of the cave entrance indicated that many meters of excavated breccia was dumped in front of the cave, much of which has revegetated over the decades and is now covered by bushes (**Figure 8.10**). Two pathways provided clear areas for the radar antenna transects.

A reflection profile was collected from a small terrace, which is bounded by a shepherd's wall, downslope from the entrance of the cave. The antennae were moved up a trail during data collection to just outside the cave entrance (**Figure 8.10**). A small out-crop of the limestone bedrock was exposed just below the beginning of the line, which allowed the distinct planar reflections visible in the profile to be tied directly to that sur-face below the limestone rubble (**Figure 8.11**). After the reflection profile was corrected for topography the bedrock reflection is visible as a horizontal high-amplitude reflection that is progressively buried more breccia toward the cave entrance. This generally horizontal surface is no longer visible when the overburden exceeds about 3 m, but still indicating that there was a horizontal land surface directly in front of the cave entrance in the past. Within about 4–6 m from the cave entrance a vertical discontinuity between

Figure 8.10 Location of the GPR profile testing the thickness of back-fill and the buried landscape in front of the Lapa do Picareiro cave, Portugal.

bedrock and the excavated breccia indicates that there was a dramatic cliff directly under the cave opening prior to the excavations. The remnant of the shepherd's wall may also be located in this area, but could not be differentiated from the bedrock margin.

The material excavated from within the cave and dumped off the cliff beginning in the 1950s, exhibits a jumbled and confusing sequence of reflection surfaces in the reflection profile (**Figure 8.11**). It is interesting that the bedrock rubble that makes up the fill inside the cave allowed 270 MHz energy to penetrate 6–8 m but outside this same material allows only 3 m of energy penetration. This difference is probably related to moisture retention in this breccia from precipitation outdoors and perhaps the addition of clay over time, both of which would raise the electrical conductivity of this ground, decreasing energy penetration.

The vertical discontinuity between the bedrock and the excavated breccia is visible by analyzing the geometry of the reflections in the profile (**Figure 8.11**). Well-layered reflections are visible from the intact bedrock units along what was the cliff. These contrast with the jumbled reflections from the waste breccia downslope (**Figure 8.11** enlarged area). The bedrock surface then becomes horizontal again, indicating that there was a small flat terrace in front of the cave entrance during the long period of time it was occupied. This GPR analysis shows that if there was a prehistoric discard area that could still be tested, it is along the base of the cliff under more than 4 m of breccia waste material.

In highland Ethiopia a small cave named Tuwaty was tested with GPR to study the floor stratigraphy that might hold evidence of early agriculture in this area. The cave is a weathered lava tube within a thick sequence of basalt flows (**Figure 8.12**). A grid of 400 MHz reflections was collected on the generally flat floor of the cave with profiles showing energy penetration only to about 80 cm (**Figure 8.13**). A high-amplitude planar reflection between 60 and 70 cm appears to be the deepest coherent reflection, suggesting the presence of a layer there that is composed of some material that attenuates radar energy. A test excavation exposed that horizon (**Figure 8.14**), which was a 5 cm thick white phosphorus deposit that is probably bat guano. The salts in this material produced a very electrically conductive medium, precluding radar energy penetration below it.

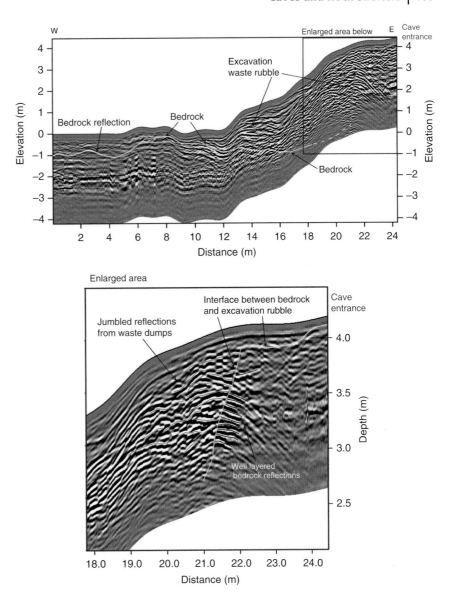

Figure 8.11 Topographically adjusted 270 MHz reflection profile showing the limestone breccia discard material in front of the cave entrance, with the buried bedrock visible as a distinct, almost vertical interface just in front of the entrance.

The sediment layers overlying the bat guano unit are generally horizontal with minor undulations between the front and rear of the cave (**Figure 8.13**). The 400 MHz antennae used here did not produce profiles with much resolution, but close examination shows a number of subtle bowl-shaped reflections within the sequence that may be hearths. They are correlative to a unit in one excavation that contained some charcoal ash, but no hearths (**Figure 8.14**). Perhaps these bowl-shaped depressions may be some type of cave floor feature other than hearths, or the hearths have just not been tested. One hypothesis is that the cave was occupied early, and then usage ceased for some reason, allowing the bats to recolonize the cave, producing the guano layer (**Figure 8.14**). People then began to use it again, driving the bats out.

Cave entrance

Collecting 400 MHz
profiles on cave floor

Figure 8.12 Collecting 400 MHz
reflection profiles inside the Tuwaty
Cave, Ethiopia.

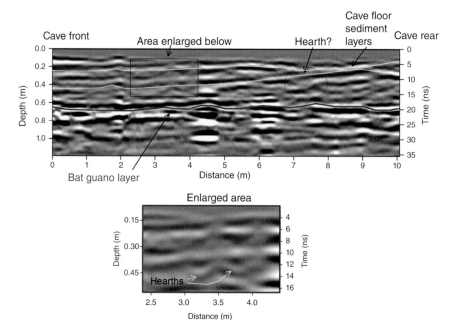

Figure 8.13 A 400 MHz
reflection profile within the
Tuwaty Cave, Ethiopia, showing
the attenuating bat guano layer
and subtle low-amplitude
reflections from hearths within
the overlying cave strata.

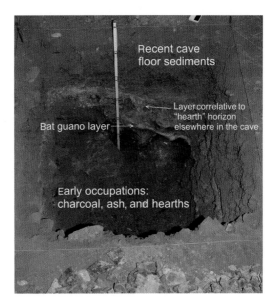

Figure 8.14 Stratigraphy in Tuwaty Cave, Ethiopia, showing dark ash and charcoal horizon from the early occupations with the white phosphorus-rich bat guano layer, overlain by later occupational units. Photograph courtesy of Matthew Curtis and John Arthur.

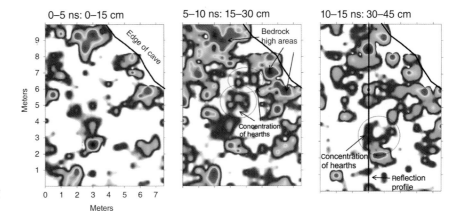

Figure 8.15 Amplitude slice-maps of the cave floor sediments at Tuwaty Cave, Ethiopia. Conyers (2012). Reproduced with permission of Left Coast Press, Inc.

Some of the reflection profiles show large blocks of bedrock at shallow depths in the cave, some of which outcropped at the surface and were visible during collection. When all the profiles were sliced into 15 cm layers, the areas of bedrock appear as high-amplitude reflections, and the small bowl-shaped features (**Figure 8.13**), which may be hearths or some other type of feature, are clustered between these bedrock projections (**Figure 8.15**).

References

Bicho, Nuno, Haws, Jonathan, & Hockett, Bryan (2006) Two sides of the same coin—rocks, bones and site function of Picareiro Cave, central Portugal. *Journal of Anthropological Archaeology*, vol. 25, no. 4, pp. 485–99.

Conyers, Lawrence B. (2012) *Interpreting Ground-penetrating Radar for Archaeology*. Left Coast Press, Walnut Creek, California.

Gilmore, Kevin P. (2008) Ritual landscapes, population, and changing sense of place during the Late Prehistoric transition in Eastern Colorado. In: Bonnie J. Clark & Laura L. Scheiber (eds.) *Archaeological Landscapes on the High Plains*, pp. 71–114. University of Colorado Press, Boulder, Colorado.

Goebel, Ted, Graf, K.E., Hockett, Bryan, & Rhode, David (2003) Late-Pleistocene humans at Bonneville Estates Rockshelter, eastern Nevada. *Current Research in the Pleistocene*, vol. 20, pp. 20–3.

Goldberg, Paul & Macphail, Richard I. (2008) *Practical and Theoretical Geoarchaeology*. Blackwell Publishing, Oxford.

Haws, Jonathan A. (2012) Paleolithic socionatural relationships during MIS 3 and 2 in central Portugal. *Quaternary International*, vol. 264, pp. 61–77.

Murphy, Laura R. & Mandel, Rolfe D. (2012) Geoarchaeology and paleoenvironmental context of the Burntwood Creek Rockshelter, High Plains of northwestern Kansas, USA. *Geoarchaeology*, vol. 27, no. 4, pp. 344–62.

Tchakirides, Tiffany Forbes (2002) *Ground-penetrating Radar and Archaeology: An Integrative Approach for Studying the Cherry Creek Canyon Rock Shelter*. Unpublished master's thesis, University of Denver, Denver, Colorado.

Wallis, Lynley A., Keys, Ben, Moffat, Ian, & Fallon, Stewart (2009) Gledswood Shelter 1: initial radiocarbon dates from a Pleistocene aged rockshelter site in northwest Queensland. *Australian Archaeology*, vol. 69, pp. 71–4.

Woodward, Jamie C. & Goldberg, Paul (2001) The sedimentary records in Mediterranean rockshelters and caves: archives of environmental change. *Geoarchaeology*, vol. 16, no. 4, pp. 327–54.

9

Anthropogenic Features and Urban Environments

Abstract: Humans can be very active agents of deposition and modifiers of landscapes and therefore anthropogenic strata must be studied in a geoarchaeological context with GPR using the same techniques as other depositional environments. Shell mounds that were modified and expanded over time, producing complex layered environments can be studied using GPR profile analysis, showing depositional layers of shells as well as interbedded floors and other constructed surfaces. Natural landscapes that were filled with construction rubble to flatten surfaces for the construction of monumental architecture are visible in profiles and amplitude maps. Buried living surfaces, fill material, and new living surfaces stacked in complex stratigraphic packages can all be seen in GPR images. Where architectural features have been eroded and redeposited, the complexity of radar reflections from intact and secondary units can be difficult to differentiate, but when the origins of reflections is understood, the GPR images make a great deal of interpretive sense. In urban environments many layers of construction, renovation, and fill can be identified, and areas where archaeological materials remain undisturbed can be mapped.

Keywords: anthropogenic, middens, urban environment, floors, living surfaces, fill layers, adobe melt, architectural features

Introduction

Humans can import and generate sediments composed of many types of materials during intensive habitation or other activities that generate layers within complex anthropogenic features such as mounds and middens. These sites, and other anthropogenically generated deposits can include occupational surface layers, which when stacked over time can produce very complex stratigraphic sequences. Examples of these anthropogenic structures are Middle Eastern tells (Butzer et al. 2013; Goldberg and Macphail 2006; Matthews et al. 1997; Rosen 1986), which are similar to urban mound features in Asia and North and South America (Schilling 2012). These deposits of thick multiple-layer habitations often contain the history of many cultures over a long period of time, many of which are eroded and truncated by natural and human forces, so that only by studying the archaeologcial record in a three-dimensional stratigraphic way can these be analyzed.

Mounds of many other sorts can be the product of periodic architectural construction and renovation or just accumulations of trash that were modified for human use

Ground-penetrating Radar for Geoarchaeology, First Edition. Lawrence B. Conyers.
© 2016 John Wiley & Sons, Ltd. Published 2016 by John Wiley & Sons, Ltd.

(Stein 1992). Other types of mounds are accumulations of both trash and architectural materials, which can also result in complicated stratigraphic layering. All these types of depositional and habitation sequences can be potentially mapped using GPR, especially when integrated with excavation data.

Urban archaeology using GPR can be especially challenging as modern utilities and modifications often cut through layers and thick fill units are deposited on top of historic features of interest. Over time as older structures are demolished, their building materials are recycled, and the resulting stratigraphic sequences can be quite difficult to interpret. Floors and road surfaces can be truncated, new buildings placed on top, with different orientations, and the original functions of buildings difficult to discern. These characteristics of urban and anthropogenically influenced sites make an integration of excavation stratigraphy with three-dimensional GPR mapping especially applicable.

Middens

In northern Europe shellfish were a basic food resource, especially in Mesolithic time, where large middens are preserved (Andersen 2007; Gutiérrez-Zugasti et al. 2011). Perhaps due to ecological constraints, freshwater and marine mollusks are not as abundant in the eastern Baltic Sea area and only a few shell middens have been discovered there (Berzins et al. 2014). The only freshwater shell midden known in this area is located in Latvia, which was first excavated in 1873 (Sievers 1874 and 1875 cited in Berzins et al. 2014). In this deposit, along the margin of a freshwater lake a 1.1 m thick sequence of shells, animal bones, pottery, and bone artifacts was uncovered along with at least four human burials. Much of the site was excavated and presumed destroyed in the late 19th century.

Research was renewed at the site in 2009 (Berzins et al. 2014) after animal bones and artifacts were discovered adjacent to the site in the bed of the river, suggesting that there might still be intact deposits to be studied. A number of GPR profiles were collected over the area considered prospective for finding intact midden stratigraphy, all under a 20th century top soil. Some of the profiles showed only disturbed and homogenized materials, which is consistent with the disturbance from the late 1800s excavations. Some of the profiles, however, displayed high-amplitude horizontal strata (**Figure 9.1**), which were thought likely intact layered midden units as described in the excavations from more than a century ago. In 2011 excavations were placed in areas where the GPR profile analysis showed likely layered midden units. Mussel shell and fish bones with numerous artifacts were uncovered in 13 stratigraphic units in the portion of the midden not destroyed. Those excavations confirmed that these high-amplitude well-stratified units visible with GPR are from the prehistoric shell midden, with the surrounding low-amplitude reflection area being homogenized, and disturbed midden material from the excavations in the late 19th century, mixed with soil.

The intact units have been dated to as long ago as 5800 BP during the middle Neolithic (Meadows et al. 2014) with many pottery sherds, bone tools, and excellently preserved shell and fish remains from aquatic resources acquired in the nearby lakes. This site was a trash disposal area associated with nearby agricultural settlements, demonstrating intensive fishing and gathering associated with horticulture at this time (Berzins et al. 2014). The GPR reflection profiles readily produce images of the horizontal layers of burned shell and fish bone-rich units, surrounded by the non-reflective homogeneously disturbed material from the previous excavations conducted more than a century ago.

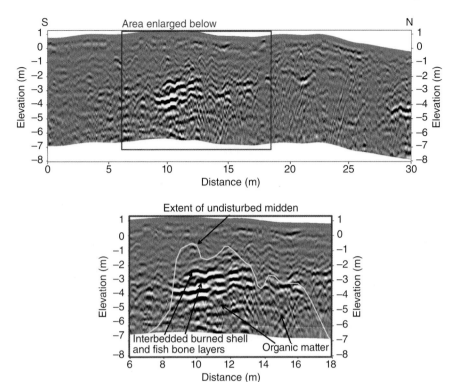

Figure 9.1 A topographically adjusted 400 MHz reflection profile collected in an attempt to find intact shell midden layers below a disturbed surface soil layer, Latvia. Data from The Section of Applied Geophysics, Institute of Geoscience, Kiel University in commission by the ZBSA, Drs. B. V. Eriksen, Harald Luebke and Christina Klein.

Along the edge of the Tagus River estuary in Portugal the Muge shell midden has been an important location for Mesolithic excavations for about a century (Bicho 1994). About 8000 years ago the Atlantic coast of Portugal became less productive as sea level rose and there was a decrease in ocean upwelling and a corresponding decline in marine resources (Bicho et al. 2010). At that time, people shifted their settlements to the Tagus Valley estuary, where the river valley had been recently inundated by marine water and the mixing of fresh and salt water produced a very productive environment for shell fish gathering (Bicho et al. 2010).

One large midden remains at the site with intact shells, the other half having been excavated periodically and removed since the 1920s. Research projects conducted on these large shell middens indicate that the initial phases of shell deposition started about 8100 BP on a fluvial terrace above an estuary just to the west (van der Schriek et al. 2007). That estuary contained very productive shell beds and people settled near here in long-term villages because of the nearby resource-rich environment. These sites were abandoned about 800 years later when the estuary started to fill with sediment, making it less productive. This was soon followed by the appearance of early Neolithic agricultural people and the displacement or assimilation of the Mesolithic people.

The largest of the partially intact shell middens was used as a test for GPR, where reflection profiles could be tied directly to excavations and the stratigraphy was visible in profile (**Figure 9.2**). The 270 MHz antennas were used for data collection and all profiles were surveyed and corrected for topography. A number of reflection profiles were collected beginning at the top of the mound, on to the surrounding flat ground, which is

Figure 9.2 Location of one of the 270 MHz reflection profiles from the top of the Muge shell mound, Portugal.

Figure 9.3 The sharp contact between the preshell living surface and the overlying shells deposited in the midden. The GPR profiles were tied directly to this excavation in order to correlate GPR reflections to known units.

the fluvial terrace tread located just above the present day Tagus River floodplain (that was the estuary in Mesolithic times). The distinct contact between the shell-rich layers of the large mound and the underlying premound living surface was visible, and all profiles were tied directly to that stratigraphy (**Figure 9.3**).

Good energy penetration occurred in the shells with a high-amplitude planar reflection generated from the preshell living surface (**Figure 9.4**). This surface represents the fluvial terrace tread above the Tagus estuary when Mesolithic people began to mound shells in this area. In one reflection profile directly beneath the crest of the mound two small mounds are visible from those initial shell accumulations. Over time shell units can be seen expanding in the mound center, sloping and thinning outward as anthropogenic deposition continued. Those shell units appear to have been deposited episodically and are bounded by soil units or finer grained depositional episodes that reflect higher amplitude radar energy along continuous planar surfaces. Toward the top of the shell sequence a very high-amplitude flat reflection is visible that might be a prepared surface constructed to level and prepare the floor for some reason, perhaps an event associated with feasting or other group activities. Shell deposition again occurred, with about 1.5 m of additional shells places on the mound top, and further spread out onto the ground surrounding the mound. There are a few high-amplitude reflections within this uppermost unit. Small hyperbolic reflections near the present ground surface all over the mound are reflections from tree roots and animal burrows.

Another profile on this shell mound (**Figure 9.5**) displays the same preshell surface, with similar shell mounding episodes. As only five test profiles were collected on this mound, no three-dimensional analysis of the shell mounding history was possible but could be studied in the future with a closely spaced grid of profiles. The reflection profile in **Figure 9.5** also displays a Pleistocene fluvial channel consisting of many high-amplitude reflections generated from its gravel and sand interfaces. Those sediments are today preserved below the terrace tread, on which the Mesolithic shells were deposited.

The GPR profiles in this shell mound can be used to study mounding episodes, modification of the shells over time to produce anthropogenic surfaces near the top of the mound, and other possible features within it. There are also many hyperbolic and planar reflections visible directly on or just beneath the premound living surface, which could be important in understanding activities in this area before shells were deposited. Many human burials have been excavated within the pre-shell sediment and could potentially be visible with GPR analyses, but this was not done in this study.

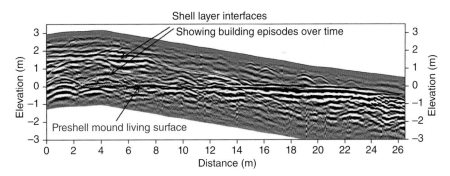

Figure 9.4 A 270 MHz reflection profile collected on the top and flank of the Muge shell midden, Portugal. The pre-shell living surface is horizontal and shell layers deposited on it are visible as moderate amplitude reflections.

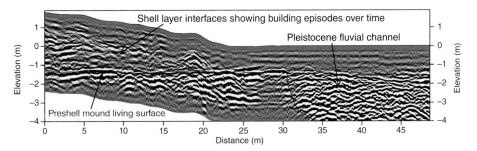

Figure 9.5 A Pleistocene fluvial channel and the preshell mound living surface on which shells were deposited is overlain by shell layers in this 270 MHz reflection profile from Muge, Portugal.

Anthropogenic Deposits

At Petra, Jordan, the late Nabatean and Roman architectural monuments and buildings are world famous for cut-sandstone facades constructed along cliff faces as well as the standing architecture in the valley just west of the dramatic cliffs. The monumental architecture has long been the focus of archaeological research, with temples and associated gardens (Bedal 2004; Gleason 2014; Markoe 2003) providing important information about the complex societies who built this city along historic trade routes from Arabia to the Mediterranean. Geophysical research there has focused on the near-surface monuments (Conyers et al. 2002) and water distribution systems (Urban et al. 2012), with many interesting buried finds discovered in the upper meter or two of the present day ground surface.

Amplitude slice-maps in the upper 50 cm or so of the present ground surface in one location at Petra show temples of Roman age in one area along a southern terrace bounding a commercial street (**Figure 9.6**). When the amplitude maps in the upper 75 cm of the surface are studied many other reflection features are also visible that were excavated and found to be platforms and other architectural features built within a formal garden, with other reflections of unknown origin.

An analysis of reflection profiles demonstrates that the stratigraphy is quite complex in this study area of Petra, with a distinct deep high-amplitude planar reflection surface dipping to the north, toward what was the valley bottom in early Nabatean times, centuries before Roman annexation in 64–63 BC (**Figure 9.7**). This horizon is the living surface on which people built much more humble structures during the early occupation of this area, prior to renovations and the construction of monumental buildings (Conyers 2010; Conyers and Leckebusch 2010). This oldest level has been uncovered only in one small excavation just to the north of the GPR study area where the surface was excavated and dated by pottery associated with it (Conyers 2010).

Prior to building the significant structures visible with GPR near the surface, or visible as still-standing buildings, the study area was leveled and the sloping ground surface was covered with rubble (**Figure 9.6**). This jumbled high-amplitude material is visible in the amplitude maps as linear fill units, probably composed of stone construction material that was distributed near where older buildings stood along possible pathways or roads (Conyers 2010). These stone piles of fill are visible in profiles as high-amplitude mounded deposits parallel to the valley margin with intervening non-reflective sediment, which is likely sand or other fill that was incorporated with the architectural rubble (**Figure 9.7**). Level surfaces that are likely pavements or compacted ground from the later Roman period can be seen above the fill layers.

At this site the GPR profiles and amplitude maps provide a way to view this site in three dimensions and understand how the ancient landscape was modified for monumental construction. This was done by destroying many stone structures and placing fill in the valley that was composed of architectural materials scavenged from those structures. The linear arrangement of these fill units is likely showing the general arrangement of the buildings before the landscape renovation, aligned with pathways and the natural stream valley. Many interesting reflections are visible directly on this buried living surface, which have not been excavated, but are likely architectural remnants of the earliest inhabitants of Petra.

On a high Pleistocene river terrace elevated above the active floodplain of a river in southern Arizona a classic period Hohokam site is today visible as a large mound (**Figure 9.8**). Many of these have been excavated and are found to contain the remains of

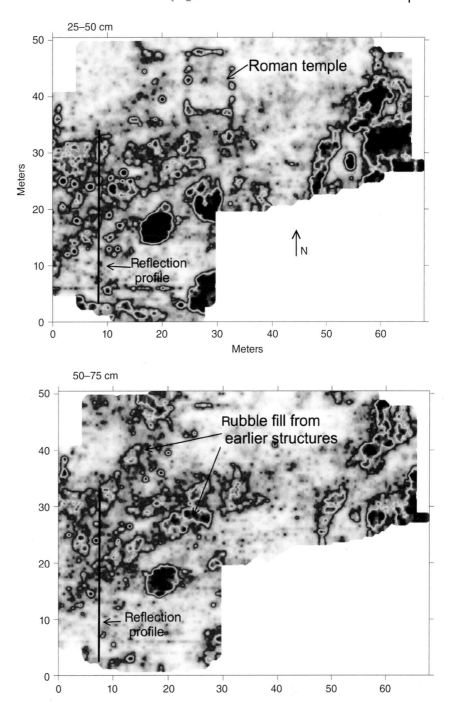

Figure 9.6 Amplitude slice-maps near the surface from an area of monumental architecture at Petra, Jordan, showing the foundations of a Roman temple and other more enigmatic reflections in the deeper slice.

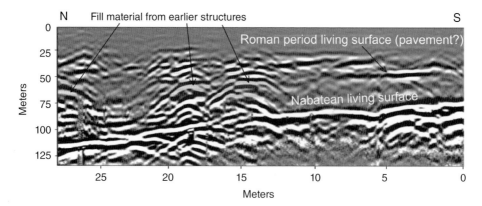

Figure 9.7 A 400 MHz reflection profile from Petra, Jordan, which shows the Nabatean living surface prior to a landscape modification that leveled the ground surface to provide a surface for later monumental construction activity. Location of this profile segment is shown in **Figure 9.6**.

Figure 9.8 An eroded adobe mound from a classic period Hohokam multistory dwelling that is today a mixture of eroded clay, some intact floors, and the bases of walls, with many associated artifacts.

clay floors, walls, and artifacts, encased by layers of "adobe melt," which are redeposited clay units derived from eroded architecture. The only other active depositional agents are a minor amount of aeolian sand interbedded with many interbedded units of clay derived from wall and ceiling collapse. There is a minor amount of slumping as the structures continued to erode with every rain storm over centuries.

This geoarchaeological setting consists of some buried architecture surrounded by a thick matrix of sediment consisting of eroded architecture, both of which are composed of the same material. The only externally derived materials are very minor laminae of aeolian sand. The similar composition of all these components within the mounds presents a challenge for GPR, as it is differences in materials along interfaces that produce radar energy reflections, and here they are composed of almost exactly the same substance.

These large multiroom pueblos in southern Arizona were constructed during a time of increasing social complexity when people moved into integrated communities that were often walled and had large elite dwellings within those walls (Craig et al. 1998; Fish and Fish 2007; Pailes 2014). The rooms were often renovated many times, buildings added on to, and sometimes burned during possible termination rituals at the death of important people, or as general "house cleaning" activities, once abandoned.

The walls of these dwellings were composed of compacted homogeneous soil and sediment that was locally obtained, mixed with water and bound by sand and gravel (Conyers 2011). This cohesive mixture was "puddled" periodically in one location and then sculpted to create walls, which dried in the hot desert sun to produce sturdy walls, sometimes many stories tall. Floors of upper stories and roofs were composed of wood and thatch. The carbonate in the soil produced a "cement-like" additive, which helped keep the walls intact, as long as rain water was shed by the roofs and channeled away from the structures.

The mixing process that was used to produce the wall material created a homogeneous clay, sand, and gravel mixture with few interfaces that can reflect radar waves. In addition, the remaining vertical portions of un-eroded walls are mostly oriented parallel to the direction of radar travel into the ground from the surface antennae, and do not therefore provide a surface from which to reflect energy. The location of the walls are only visible using GPR by mapping the adobe melt beds interbedded with aeolian sand, which were deposited as they were eroded on the flanks of walls (Figure 9.9).

Compacted earthen or clay floors produce distinct high-amplitude planar reflections in profiles and map view (Figure 9.10). The melt layers adjacent to the still-intact walls also produce high-amplitude reflections. Numerous excavations have uncovered these melt beds, demonstrating that the high-amplitude reflections were generated from interfaces between layers of melt and thin sandy aeolian beds deposited by wind deposition between the rain storms that deposited the adobe.

Once the origin of reflections was understood (Conyers 2012a), the amplitude slice-maps could be interpreted, with a focus on the areas of no-reflection, as those are the locations of still intact walls composed of homogeneous mud and aggregate (Figure 9.11). The high-amplitude areas of the maps are where the interbedded adobe melt and aeolian sand in-filled rooms with partially eroded walls, or along the outside margins of those walls. The only high-amplitude point-source reflections visible in the slice-maps were from large rocks, some of which were grinding stones left in the corners of some rooms (Figure 9.9).

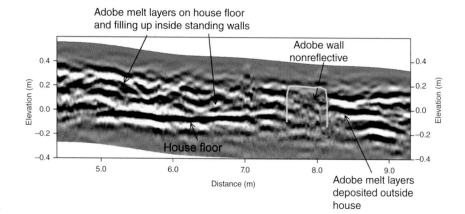

Figure 9.9 Adobe melt layers adjacent to walls were deposited after the structures were abandoned, producing reflective surfaces for radar energy. The only individual artifacts in these mounds that are reflective are large rocks, such as the grinding stone in the inset picture.

Figure 9.10 Four hundred megahertz reflection profile showing high-amplitude planar house floor reflections with an adjacent wall that is non-reflective. Adobe melt layers on top of the floor and adjacent to the walls provide high-amplitude reflections that denote the location of the walls.

Urban Settings

Some of the most complicated environments for geoarchaeology and GPR are in urban areas that have been subject to multiple periods of construction, demolition, rebuilding, and excavation. One notable example of this type of GPR study is from Leicester, England, where the recently uncovered remains of King Richard III were discovered in a parking lot under many layers of sediment (Buckley et al. 2013). His skeleton was found under the floor of the Medieval friary of Gray Friars, which was built between the years 1224 and 1230.

Ground-penetrating radar data were acquired and interpreted in the parking lot prior to excavations and the method was deemed a failure as only modern utilities were visible (Austrums 2011, cited in Buckley et al. 2013). The body of King Richard III was discovered in 2013 after trenching in the parking lot that began in 2012. Prior to locating his remains, the walls of a chapter house and a possible cloister garth (a garth is an open space within the confines of a friary) were uncovered, and then later the friary church.

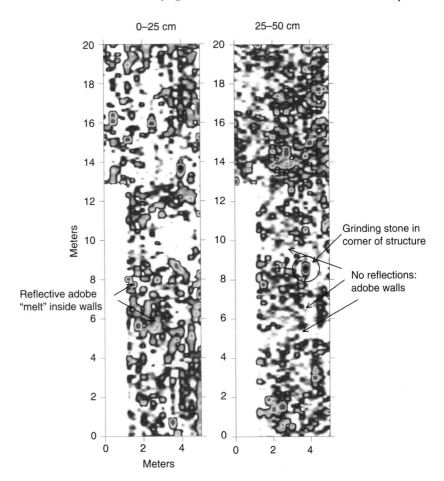

Figure 9.11 Amplitude slice-maps of a two-room adobe structure showing non-reflective walls and adjacent adobe melt layers within rooms, and on the flanks of structures outside the eroded walls.

In one trench within the walls of the presbytery (the portion of the church reserved for the clergy), below the floor, the remains of King Richard III were discovered.

This king of England was killed at the Battle of Bosworth at the termination of the War of the Roses on August 22, 1485, and his body, according to legend, was buried in the floor of this church. The structure was demolished sometime after 1538 and the area became an urban garden, a dump for debris from nearby mills, and later part of urban downtown Leicester. Most of the buildings in this area of town were built beginning in the 18th century, with modern sewer, water, and other utilities installed in the last century. The remains of the church and associated buildings were vaguely known from historic maps and documents, and therefore excavations were initially conducted in the only available area (the parking lot) in a way that would most likely encounter foundations, if they existed (Buckley et al. 2013). Those excavations revealed the foundations of the Grey Friars Church and associated buildings, confirming their general location and portions of foundations, floors, and a general outline of the original structures. The lack of good building stone in this area of England no doubt led to a recycling of all good stones from the demolished structures during Medieval time, which demonstrates why the initial GPR survey had such difficulty producing images of much more than the fill and utility pipes that overlay the church remains of interest.

A GPR grid was collected in 2013 using 400 MHz antennae in a small parking area just west of the ongoing excavations that uncovered the presbytery (**Figure 9.12**). Another much larger grid of profiles was also collected in a large parking lot to the south, the results of which are not discussed here, as the area had been highly disturbed by recent construction. The stratigraphy adjacent to the GPR grid was visible in a number of trench walls and the exposed foundations of the church and other architectural remains were also visible, which could be projected into the subsurface (**Figure 9.13**).

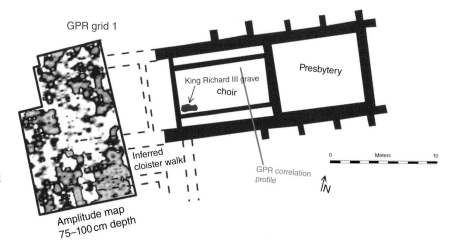

Figure 9.12 Location of the presbytery and the King Richard III grave, as exposed and published by Buckley et al. (2013), with the location of GPR grid 1. The location of one correlation reflection profile used to directly tie GPR reflections with known stratigraphy is shown in the middle of the church.

Figure 9.13 The exposed presbytery foundation and portions of the walls that are composed of sandstone and could be projected into the areas still preserved under the parking lot surface and tied to a GPR profile.

The general stratigraphy visible in the excavation walls shows the Medieval friary level at about 90–120 cm depth below the surface of the parking lot. The architectural remains consist mostly of foundations and a few courses of wall stone, covered by soil and smaller blocks of rubble that remained here after the friary stones were scavenged to be used elsewhere. It appears that the friary walls may have remained standing for some years after abandonment, as a weak soil layer accumulated on the floors of the church (Buckley et al. 2013). Soil then appears to have formed again on all the surrounding friary debris, with no evidence from the excavations that the area was built on in this location (**Figure 9.14**). Mixed with the soil at a time this area was a garden is a mixture of rubble of unknown origin.

In other locations the unit overlying the Medieval layer where the church foundations are preserved is a layer of breccia and architectural debris generally referred to as "post-Medieval rubble" (**Figure 9.15**). This unit is generally correlative to the garden soil and rubble elsewhere at the site, both of which vary in thickness from about 30–40 cm. That stratigraphic horizon is in turn overlain by fill that was used to level this area starting in the late 18th century when industrialization began in Leicester and leather, wool weaving, shoe making and, other manufacturing activities were occurring nearby. The parking lot surface directly overlays this mill waste rubble in much of the study area.

More recent intrusive features are visible throughout the site starting with cellars built in the 18th and 19th centuries and the emplacement of more recent pipes in the 20th century (**Figures 9.14** and **9.15**). Excavations cut through all the units of interest during the emplacement of these recent utilities and cellars.

One correlation profile was collected with the 400 MHz antennas along an area of the parking lot surface (**Figure 9.12**) directly adjacent to an area where the friary wall remains and foundations were exposed (**Figure 9.13**). The depths of the stratigraphic units, all of which are visible in the trench, were then directly correlated to the reflections visible in the GPR profile (**Figure 9.16**). The highest amplitude reflection is the contact at the base of the post-Medieval fill unit with the underlying Medieval architectural horizon. Just above that contact is where the post-abandonment soil formed in some locations, often mixed

Figure 9.14 The darker garden soil is mixed with Medieval architectural rubble overlying the remains of the church where the body of Richard III was found. That unit is overlain by mill waste rubble from nearby industries that disposed of materials starting in the 18th century. A modern cellar intrudes into all these units.

Recent utilities and 19th–20th century intrusive features

Figure 9.15 The three basic stratigraphic units at the Grey Friars church site are visible in this profile with the Medieval level on which the church was built in the 13th century. It is overlain by a post-Medieval rubble layer (sometimes also including the garden soil visible in **Figure 9.14**) that was deposited or formed after the church was destroyed in 1538. Rubble from a nearby industrial mill was disposed of in this location starting in the 18th century, all capped by the modern parking lot surface.

Figure 9.16 A 400 MHz reflection profile used for correlation of the stratigraphy visible in a trench wall with the units displayed in the profile.

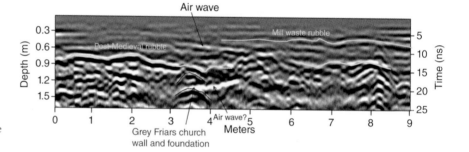

with waste rubble from the destroyed structures. The remains of the Grey Friars Church are visible as high-amplitude reflections that would be difficult to differentiate from the surrounding rubble if there was not a direct visible correlation available in the adjacent trench (**Figure 9.16**). Two weak "**air waves**" are visible in the profile, which were recorded from waves reflected off the walls of nearby buildings. They are noticeable as straight reflections passing through the normal stratigraphic units, which confuses this reflection profile to some extent. These types of reflections are common in urban GPR, caused by radar waves moving along the ground surface and in the air from the transmitting antenna and then back to the receiving antenna in the air (Conyers 2012b, p. 84; 2013, p. 85).

Changes in the post-Medieval rubble layer laterally produce a variation in the amplitude of waves generated at its interface with the underlying architectural rubble from the Grey Friars Church (**Figure 9.16**). Moisture differences in the overburden layers also created velocity variations that make this mostly horizontal interface appear to vary with depth. Along just a 9 m long profile these variations are considerable. A very weak reflection is also recorded from the boundary between the mill waste rubble unit and the underlying post-Medieval rubble (**Figure 9.16**). These two units are not very different in composition, so the interface between them is not visible as a high-amplitude reflection in most of the profiles in this area. Many other reflections are visible in the stratigraphic horizon where the Grey Friars Church wall is visible, whose origin is not known. These were likely generated from a variety of architectural rubble that remained after the destruction of the church.

In the small grid to the west of the Grey Friars church a very complex urban stratigraphy is preserved under the parking lot (**Figure 9.12**). Called grid 1, this 10 × 15 m area shows how complicated these types of environments are for GPR, but also how they can be understood by studying and comparing both the profiles and the amplitude maps. Each of the reflection profiles, which were 50 cm apart, had to be interpreted individually starting with tying all horizons directly to the profile that was correlated directly to the exposure (**Figure 9.16**). This type of manual interpretation is necessary in complex urban ground of this sort, prior to trying to interpret reflection amplitudes in map view (Conyers 2012b, p. 28).

The profile 1.5 m east of the edge of grid 1 (**Figure 9.17**) shows the high-amplitude planar reflection at the base of the post-Medieval fill unit (**Figure 9.18**). This undulating surface is exactly the same as in the correlation profile (**Figure 9.16**) but about 20 cm shallower than in the correlation line, which is located 20 m to the east.

The base of the post-Medieval rubble layer is truncated in the middle of the profile by a recent pit, filled with rubble (**Figure 9.18**). The rubble clasts all generate individual reflections, producing a very busy unit of large individual reflections. The rubble layers overlying this interface, which are composed of smaller clasts, are not large enough to produce reflections with the 400 MHz antenna energy. These appear in the reflection profile to be a mostly homogeneous rubble layer. A few stratigraphic horizons in that uppermost unit may be contacts between rubble horizons, such as the one barely visible in **Figure 9.15** at the base of the

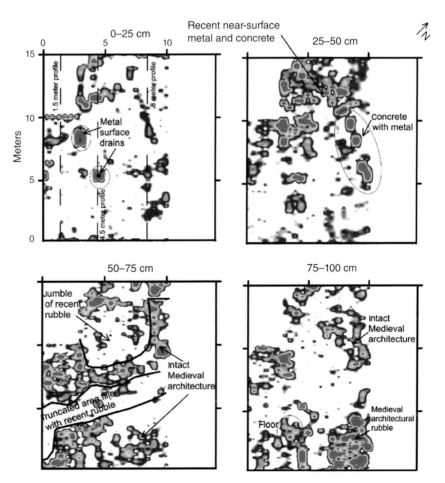

Figure 9.17 Amplitude slice-maps of GPR grid 1 at Grey Friars Church, Leicester, England, with the annotations of important features visible in profiles and correlated with reflections in map view. The location of three profiles used as illustrations of stratigraphy are shown in the 0–25 cm slice.

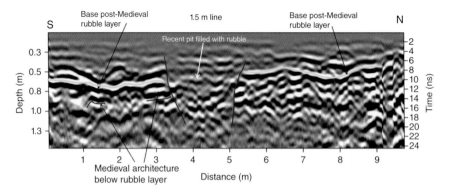

Figure 9.18 Reflection profile 1.5 m from the west line of GPR grid 1 showing the undulating reflection at the base of the post-Medieval rubble with a few high-amplitude reflections at the depth where Grey Friars architectural features should be found. This area is truncated by a trench a little more than a meter wide, the excavation of which has partially removed the stratigraphy of interest.

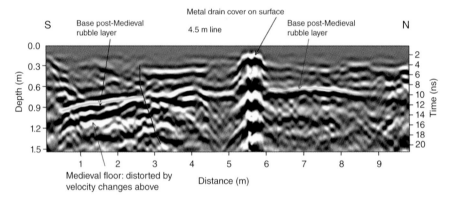

Figure 9.19 A 400 MHz reflection profile displaying the surface metal drain cover that is directly on top of a modern drainage pipe. The base of the post-Medieval rubble layer overlays a floor on the south side of the profile. The middle of the profile shows truncation by a large excavation, perhaps related to the insertion of the drainage pipe.

mill rubble layer. Some reflections that are characteristic of those from the known church foundations are visible below the base of the post-Medieval rubble (**Figure 9.18**). These may be associated with walls around an inferred garth or cloister walk near the church (Buckley et al. 2013). The various rubble reflections on the north end of the profile (**Figure 9.18**) beneath the post-Medieval rubble horizon produce a very busy incoherent group of reflections. Perhaps that area only contains architectural pieces that were used as fill.

Just 3 m farther east from Profile 1.5 a very different stratigraphic profile is apparent (**Figure 9.19**). This profile shows the same base post-Medieval rubble horizon, directly on a planar reflection, which is likely a Medieval floor or pavement surface. That surface is not perfectly horizontal, probably because of minor variations in water saturation in the overlying material (Conyers 2013, p. 158). The stratigraphy here is also truncated by an excavation that is more than 3 m wide in this area of the grid. This appears to connect to the narrower incision visible in **Figure 9.18**. A surface metal drain cover produced a distinct surface metal reflection. This drain cover likely connects to a metal pipe that was emplaced in this trench in recent times. The post-Medieval rubble horizon is visible on the north portion of the profile, with no distinct high-amplitude reflections from any intact

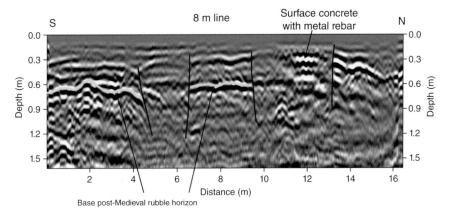

Figure 9.20 Reflection profile on the eastern side of the GPR grid displaying two areas of incision and removal of the post-Medieval rubble horizon and all other materials stratigraphically below it. A buried piece of concrete with metal reinforcing bars or wire mesh is visible at 12 m.

architecture. A distinct horizontal layer of fill is visible as a high-amplitude reflection, which defines the southern edge of the modern disturbance about 30 cm below the ground surface (**Figure 9.19**). This unit has no archaeological importance other than to help in the interpretation of the recent excavations.

Close to the eastern edge of the GPR grid (**Figure 9.20**) the same base-Medieval fill horizon is visible, but here the excavations that cut through it show two distinct trenches. No intact architectural materials are visible below this horizon, although many reflections are visible that may be rubble associated with the destruction of the Grey Friars church. Within the trench area on the north a modern horizontal layer of concrete with metal reinforcing bars or wire mesh is visible (Conyers 2012b, p. 83).

The complexity of this small grid of GPR data shows why the initial GPR survey in this parking lot appeared to show only modern utilities. No excavations were available at that time, which would allow radar reflections visible in profiles to be correlated to known stratigraphy. Amplitude slice-maps showing the distribution of buried features would be overwhelmed by linear pipes, excavation trenches, and numerous other incised features from cellars. Only after stratigraphy was known from excavations and visible in GPR reflection profiles could each of the profiles be interpreted individually. Then a spatial analysis of stratigraphy and associated archaeological materials within this complex urban area could be made.

A study of the reflections from 25 cm thick levels in the amplitude slice-maps (**Figure 9.16**) shows that down to about 50 cm in this area only metal, architectural debris from post-Medieval times, or trench fill and other waste materials are found. The areas of truncation are visible as areas of no reflection in the 50–75 cm depth slice, as the sediment that filled these excavations was mostly homogenized and contained no intact architectural pieces. Adjacent to the excavated areas many reflection features display Medieval architecture that may have been only partially disturbed during the destruction of the Grey Friars buildings, and these remain largely undisturbed since that time. The floor visible below the post-Medieval rubble fill layer (**Figure 9.19**) is only preserved in a small undisturbed area in the southwestern portion of the grid.

In this complex area of urban geoarchaeology more than 80% of the volume of ground that was imaged with GPR consists of overburden or is trench fill and other highly disturbed mixture of materials. The architecture of interest remains, but only in a few small areas. Any amplitude slice-mapping of this grid without an analysis of each individual reflection profile would generate an extremely difficult array of reflections that would preclude any coherent analysis.

References

Andersen, Søren H. (2007) Shell middens ("Køkkenmøddinger") in Danish prehistory as a reflection of the marine environment. In: Nicky Milner, Craig E. Oliver, & Geoffrey N. Bailey (eds.) *Shell Middens in Atlantic Europe*, pp. 31–45. Oxbow Books Ltd, Oxford.

Bedal, Leigh-Ann (2004) *The Petra Pool-complex: A Hellenistic Paradeisos in the Nabataean Capital.* Gorgias Press LLC, Piscataway, New Jersey.

Berzins, Valdis, Binker, Ute, Klein, Christina et al. (2014) New research at Rinnukalns, a Neolithic freshwater shell midden in northern Latvia. *Antiquity*, vol. 88, no. 341, pp. 715–32.

Bicho, Nuno Ferreira (1994) The end of the Paleolithic and the Mesolithic in Portugal. *Current Anthropology*, vol. 35, no. 5, pp. 664–74.

Bicho, Nuno, Umbelino, Claudia, Detry, Cleia, & Telmo, Pereira (2010) The emergence of Muge Mesolithic shell middens in central Portugal and the 8200 cal yr BP cold event. *The Journal of Island and Coastal Archaeology*, vol. 5, no. 1, pp. 86–104.

Buckley, Richard, Morris, Mathew, & Appleby, Jo (2013) The king in the car park: new light on the death and burial of Richard III in the Grey Friars church, Leicester, in 1485. *Antiquity*, vol. 87, no. 336, pp. 519–38.

Butzer, Karl W., Butzer, Elisabeth, & Love, Serena (2013) Urban geoarchaeology and environmental history at the Lost City of the Pyramids, Giza: synthesis and review. *Journal of Archaeological Science*, vol. 40, no. 8, pp. 3340–66.

Conyers, Lawrence B. (2010) Ground-penetrating radar for anthropological research. *Antiquity*, vol. 84, no. 323, pp. 175–84.

Conyers, Lawrence B. (2011) Ground-penetrating radar mapping of non-reflective archaeological features. In: Mahmut Drahor & Meric Berge (eds.) *Proceedings of the 9th International Conference on Archaeological Prospection*, Sept. 19–24, Izmir, Turkey, pp. 177–9. Archaeology and Art Publications, Istanbul, Turkey.

Conyers, Lawrence B. (2012a) Advances in ground-penetrating radar exploration in southern Arizona. *Journal of Arizona Archaeology*, vol. 22, pp. 80–91.

Conyers, Lawrence B. (2012b) *Interpreting Ground-Penetrating Radar for Archaeology*. Left Coast Press, Walnut Creek, California.

Conyers, Lawrence B. (2013). *Ground-penetrating Radar for Archaeology*, 3rd Edition, Altamira Press, Rowman and Littlefield Publishers, Lantham, Maryland.

Conyers, Lawrence B. & Leckebusch, Juerg (2010) Geophysical archaeology research agendas for the future: some ground-penetrating radar examples. *Archaeological Prospection*, vol. 17, no. 2, pp. 117–23.

Conyers, Lawrence B., Ernenwein, Eileen G., & Bedal, Leigh-Ann (2002) Ground-penetrating radar discovery at Petra, Jordan. *Antiquity*, vol. 76, no. 292, pp. 339–40.

Craig, Douglas B., Holmlund, James P., & Clark, Jeffery J. (1998) Labor investment and organization in platform mound construction: a case study from the Tonto Basin of Central Arizona. *Journal of Field Archaeology*, vol. 25, no. 3, pp. 245–59.

Fish, Suzanne K. & Fish Paul R. (2007) *The Hohokam Millennium*. Publications of the School for Advanced Research, Santa Fe, New Mexico.

Gleason, Kathryn L. (2014) The landscape palaces of Herod the Great. *Near Eastern Archaeology*, vol. 77, no. 2, pp. 76–97.

Goldberg, Paul & Macphail, Richard I. (2006) *Practical and Theoretical Geoarchaeology*. Blackwell Publishing, Malden, Massachusetts.

Gutiérrez-Zugasti, Igor, Andersen, Søren H., Araújo, Ana C., et al. (2011) Shell midden research in Atlantic Europe: state of the art, research problems and perspectives for the future. *Quaternary International*, vol. 29, no. 1, pp. 70–85.

Markoe, Glenn (2003) *Petra Rediscovered: Lost City of the Nabataeans*. Thames & Hudson, New York.

Matthews, Wendy, French, Charles A., Lawrence, Thomas, et al. (1997) Microstratigraphic traces of site formation processes and human activities. *World Archaeology*, vol. 29, no. 2, pp. 281–308.

Meadows, John, Lübke, Harald, Zagorska, Ilga, et al. (2014) Potential freshwater reservoir effects in a Neolithic shell midden at Riņņukalns, Latvia. *Radiocarbon*, vol. 56, no. 2, pp. 823–32.

Pailes, Matthew (2014) Social network analysis of early classic Hohokam corporate group inequality. *American Antiquity*, vol. 79, no. 3, pp. 465–86.

Rosen, Arlene Miller (1986) *Cities of Clay: The Geoarcheology of Tells*. University of Chicago Press, Chicago, Illinois.

Schilling, Timothy (2012) Building monks mound, Cahokia, Illinois, AD 800–1400. *Journal of Field Archaeology*, vol. 37, no. 4, pp. 302–13.

van der Schriek, Tim, Passmore, David G., Stevenson, Anthony C., et al. (2007) The palaeogeography of Mesolithic settlement-subsistence and shell midden formation in the Muge valley, Lower Tagus Basin, Portugal. *The Holocene*, vol. 17, no. 3, pp. 369–85.

Stein, Julie K. (1992) The analysis of shell middens. In: Julie K. Stein (ed.) *Deciphering a Shell Midden*, pp. 1–24. Academic Press, Elsevier, Amsterdam.

Urban, Thomas M., Alcock, Susan E., & Tuttle, Christopher A. (2012) Virtual discoveries at a wonder of the world: geophysical investigations and ancient plumbing at Petra, Jordan. *Antiquity*, vol. 86, no. 331, http://www.antiquity.ac.uk/projgall/urban331/ (accessed July 15, 2015).

10 Conclusions

Abstract: Fluvial, beach, and associated aeolian environments are the most applicable areas for GPR in geoarchaeological contexts. These areas contain abundant archaeological sites and are ideal for most GPR methods. Caves are more difficult as they contain thick floor sequences of thin stratigraphic units, which necessitate low-frequency antennae for depth penetration but accompanying lower resolution. There is a bright future for GPR in lakes, swamps, and bogs due to good depth penetration, but access and collection can be more difficult. Urban environments are often the most complicated due to the complexity of anthropogenic stratigraphy. Only detailed profile analysis in a tightly spaced grid will suffice for interpretation in these areas, and only later will amplitude slice-maps be understood. GPR for geoarchaeological analysis in the future will require multidisciplinary students trained in geology, archaeology, and geophysics.

Keywords: depositional environment, landscape analysis, stratigraphy, soils, geophysical integration

Collection of Data for this Book and the Future of GPR in Geoarchaeology

In the process of writing this book I spent a great deal of time searching for published reports and articles written about projects where people had used GPR in geoarchaeological contexts. I wanted to see what others' methods showed, and how those scientists applied GPR to their archaeological projects, with particular attention to how an analysis of stratigraphy, soils, and other geoarchaeological concepts were used. What I found surprised me, as there was much research on the use of stratigraphy, soil, and geomorphology work incorporated into archaeological projects but almost nothing in those studies that also included GPR. I also found that there are many archaeologists using GPR, and the published literature is varied and robust, but again, almost nothing where all three subfields discussed in this book (archaeology, geology, and GPR) were integrated.

It was heartening to read the articles by many researchers who are using GPR in purely geological studies, which have been spawned by environmental hazard questions, climatic change analysis, and engineering applications. But in these studies, the human element may be discussed vaguely, but none of those purely geological GPR uses can be considered geoarchaeological GPR. There has also been a good deal of GPR work

Ground-penetrating Radar for Geoarchaeology, First Edition. Lawrence B. Conyers.
© 2016 John Wiley & Sons, Ltd. Published 2016 by John Wiley & Sons, Ltd.

done recently for purely archaeological purposes, and those published studies often refer to the geology at a site, but integrating geology using GPR into an archaeological framework or landscape analysis is almost never the goal. I had no idea when starting this project that an integration of all three aspects that are the focus of this book would be such a rare commodity.

In preparing examples for this book I relied on Mike Waters' book on geoarchaeology (Waters 1992) as a general guideline for a categorization of depositional environments that might have archaeological importance, and where GPR could potentially be utilized. I added to that list using the Goldberg and Macphail volume (2006), as well as Brown (1997) and Rapp and Hill (2006). In the process of compiling a list of possible GPR applications in geoarchaeology I found that there were many published articles on geoarchaeological studies where geology was integrated nicely with archaeology, but almost none of them had used GPR. I therefore had to go to the purely geological GPR literature to find examples of GPR use in my chosen environments and then searched all of my own GPR datasets from the past to use as examples. Environments within which I had no GPR examples were found in the archives of friends and acquaintances who agreed to provide me with good GPR examples that *could* have geoarchaeological applications, but where their own research focus was purely geological. Readers of this book will no doubt have noticed where these are, as I usually concluded by discussing how GPR *might be* used in these environments with the examples provided.

It is apparent that many people are using GPR, often for only geological applications, with others applying GPR in purely archaeological studies, and both dabbling with the integration of all. This reminds me of where GPR was for purely archaeological applications in 1993 when Dean Goodman and I first began collaborating (Conyers and Goodman 1997). It was a very lonely world in the GPR-archaeological community at that time for many reasons, notably the lack of digital GPR systems and software to process those data. That has now changed and there are thousands of people using GPR in archaeology all around the world. My guess is that this history of GPR use over the last 25 years is much similar to where we are today with GPR for geoarchaeology. But today, it is not the lack of hardware and software that has stifled a more general application of GPR for geoarchaeology, but perhaps a lack of people disposed to integrate all these subfields.

This shortage of practitioners of GPR for geoarchaeology is most likely a function of the lack of training and background, and not the willingness of those potential researchers. Using GPR for geoarchaeology necessitates that users be trained in stratigraphy, sedimentation, soils, and geomorphology (the geological sub-disciplines), with a good background in archaeology. Those people also need to be proficient in geophysics with a concentration on GPR. Only then can all these subfields be integrated within what are usually very complicated three-dimensional sedimentary packages of ground. Therefore, to move forward in the application of GPR to geoarchaeology there are going to have to be people trained in all these disciplines who are willing to apply multiple concepts, methods, and ideas to their research projects. The training to do this, not necessarily the willingness to try, is probably the most difficult hurdle to overcome. Hundreds of universities teach archaeology, a few tens of those may teach a geoarchaeology course, and fewer still have graduate programs focused on geoarchaeology. Even fewer offer geophysics as applied to archaeology, and there are none that I know of where all three are integrated in any coherent manner.

It may sound as if I am pessimistic regarding the future of GPR in geoarchaeology, and that is not the case. I continue to run into young scientific enthusiasts of GPR, who have strong interests in both archaeology and geology, and have gone out of their way to find the training in all three. It will be up to the younger scientists to move what appears to be a narrow subfield in new directions. In preparation for this book, I became acquainted with one such young Brazilian scholar, Tiago Attore, who had a spectacular GPR dataset with very important geoarchaeological applications (see his data in Chapter 6). I was able to convince him to let me process some of his data for this book, as I needed some near-shore aeolian examples. We spent many weeks corresponding as I processed his data (he had his own processed images using different software), and we jointly tried to figure out the geoarchaeological complexity of his project. He is one of these young GPR researchers who came to his project in Brazil from a non-scientific discipline and had to become good at GPR on his own. He had advisors and mentors from geology, geophysics, and archaeology, but it was up to him to integrate all three. It will likely take his kind of drive and perseverance in others to push GPR for geoarchaeology until the subject is taught as a subfield in geology, anthropology, or archaeology departments at universities.

Environments Where GPR is Most Applicable in Geoarchaeology

The most common environments for archaeological applications, in which "everyday" archaeologists can use GPR with geoarchaeology are in fluvial systems, river terraces, and associated environments where soil horizons are part of the stratigraphic package. Also high on the list are aeolian environments and beaches with associated dunes. For this reason, I have devoted much of this book to those environments. The GPR method is an excellent tool to study these environments and archaeological sites within these complex packages in three-dimensional ways not possible using standard geological methods.

During my background research for this book I was particularly interested in how few GPR surveys had been published from cave studies. This seems to me to be the most natural environment for GPR, but perhaps the resolution necessary for archaeologists in those contexts is beyond what can be provided by GPR. Often cave floor deposits are too deep for most GPR techniques, as with increasing depth there is a loss in bed resolution. I hope in the future people will perhaps use higher frequency antennae for high resolution in the shallow layers prior to excavations of cave floors, and then use them again on deeper and deeper levels as large-scale excavations progress downward. This might be one way to overcome the depth-resolution problem for deeply stratified caves.

The application of GPR for lake studies is also in its infancy. We now have the technology to place GPR reflection data in space, often in "real time" using GPS receivers integrated electronically with antennae. This will make boat acquisition much easier in the future, as transects could theoretically be collected by just motoring or paddling around a lake, collecting one huge dataset, which can be put into space almost immediately. Working in bogs and swamps is always going to present a challenge with the movement of antennae and people over difficult terrain. Perhaps these areas are best studied with GPR in the winter when there is ice and snow cover, or if a study area does not freeze, using of some type of shallow draft boat or other transport for swampy areas. There are undoubtedly many more very interesting archaeological bog features to be discovered and studied, such as the one presented from Scotland (Chapter 5).

Shell mounds are another natural environment for GPR, as many have high porosity, and therefore a good deal of fresh water leaching through them over time, which has removed clays and salts that would attenuate radar energy. The steep flanks of these anthropogenic features need not inhibit GPR research, as GPS systems integrated with

GPR antennae can collect topography for later profile adjustments. Low-altitude drone or kite systems can also collect photogrammetry images to produce digital elevation models (DEMs), which are very useful to correct GPR data for topography. Antennae can be outfitted with tilt meters and reflection traces can readily adjust reflection traces to the vertical in order to keep distortion of reflections at a minimum on steep-sided shell mounds (Goodman et al. 2006).

When I was writing the section on anthropogenic complexity it finally became apparent why so few people have used GPR in southern Arizona (Chapter 9). The usual complaints of the GPR method in this desert area is that the ground is full of dissolved salts, there is carbonate in the soils, and the archaeology "melts" into complex layers, making features difficult to visualize in standard GPR images. While I have discussed the adobe melt complexity in Arizona as if I understood how to overcome these problems immediately, in reality it took almost 5 years of work to understand why the non-reflective areas are what need to be understood in profiles and amplitude maps (**Figures 9.10 and 9.11**). This is not something that comes naturally to GPR people, as we are trained to analyze reflections, not areas that are non-reflective (Conyers 2011, 2012a).

Perhaps the most complex of all the environments presented here occur in urban areas where layer is placed upon layer over time, all highly modified by humans. Humans are sometimes much more difficult to figure out than geological processes, as they can act in ways not always comprehensible. Numerous cut and fill units, the burial of important anthropogenic horizons by layers of fill, and periodic scavenging of architectural components accompanied by leveling and "landscaping" are just a few of the processes that must be accounted for in these types of "very human" geoarchaeological environments. Only good correlation of GPR profiles with stratigraphy that is well understood from excavations or cores, and long and laborious work correlating the understandable units visible in GPR reflection profiles adjoining profiles within a fine-grained grid of data will work. In these types of studies amplitude maps should be produced last (or at least interpreted last), as they will almost always be so busy with reflections from this stratigraphic complexity that mistakes in interpretation will be common by relying on "bulk processing" techniques. An iterative interpretation process that compares GPR images with excavations and exposures first, with interfaces then compared between profiles and only later identified in amplitude maps is the logical interpretation progression. This is the only way to use GPR in these complexly layered urban environments (Conyers 2015).

There are many environments where GPR can be integrated with geoarchaeology, which have not been discussed in this book. I touched briefly on volcanic and volcaniclastic sediments in Chapter 3, but did not provide other examples. The use of GPR in volcanic areas perhaps deserves a book of its own, as geologists have used GPR often in lava flows and other volcanic sequences (Ettinger et al. 2014; Gomez et al. 2012). The archaeological community has only touched the surface in exploring for and mapping archaeological sites in volcaniclastic and lava flow areas (Conyers and Connell 2007). The potential use of GPR for void space mapping in volcanic areas is also an application that I have touched on elsewhere (Conyers 2012b, p. 171), and much work needs to be done with these features by GPR researchers within a geoarchaeological context. I have only dabbled with GPR in other less common or useful environments for geoarchaeology such as landslides, slump blocks, solifluction lobes, soil creep areas, and other areas of earth movement. Areas that have been tectonically altered due to faulting and fracturing might also be an interesting application for GPR (Jewell and Bristow 2006), but I know of no direct applications to archaeology on this topic. There are also many complexities that come with bedrock (Conyers 2012b, p. 65), which I have included here only in a small way (**Figures 6.5, 6.6,** and **6.7**), and this application of GPR should be part of any complete geoarchaeological package.

The Future of GPR in Geoarchaeology

Very large scale GPR surveys covering many hectares are going to be more common in the future with multiple arrays of antennae used, pulled behind ATVs and processed with very demanding and complicated software packages (Gaffney et al. 2012; Trinks et al. 2010). The processing of these huge data sets could be termed "mega-batch processing," and amazing images and videos of the ground are often produced. Even these products from huge surveys need interpretation and integration with geological and archaeological information in the form of the standard GPR image, the reflection profile (Conyers 2015).

I can see many reasons why the now narrow subfield of GPR in geoarchaeology will continue to grow, as it is a necessary addition to many research agendas. As research in archaeology focuses on landscapes and changes in landscapes because of environmental fluctuations become more common, only geoarchaeological research can provide answers in an archaeological context. In the future standard geoarchaeological analyses must be expanded over larger study areas where standard excavation, coring, or other techniques cannot provide answers. This will necessitate that researchers use GPR, as it is the only three-dimensional method that is useful in the upper few meters of the ground, which is also fast and applicable to most study areas. Growth in the application of techniques such as GPR is always spurred by necessity, and for this reason a greater future application of GPR in geoarchaeology is inevitable.

References

Brown, A.G. (1997) *Alluvial Geoarchaeology: Floodplain Archaeology and Environmental Change*. Cambridge University Press, Cambridge.

Conyers, Lawrence B. (2011) Ground-penetrating radar mapping of non-reflective archaeological features. In: Mahmut Drahor & Meric Berge (eds.) *Proceedings of the 9th International Conference on Archaeological Prospection*, Sept. 19–24, Izmir, Turkey, pp. 177–9. Archaeology and Art Publications, Istanbul, Turkey.

Conyers, Lawrence B. (2012a) Advances in ground-penetrating radar exploration in southern Arizona. *Journal of Arizona Archaeology*, vol. 22, pp. 80–91.

Conyers, Lawrence B. (2012b) *Interpreting Ground-penetrating Radar for Archaeology*. Left Coast Press, Walnut Creek, California.

Conyers, Lawrence B. (2015) Multiple GPR datasets for integrated archaeological mapping. *Near-Surface Geophysics*, vol. 13, no. 2. doi:10.3997/1873-0604.2015018

Conyers, Lawrence B. & Connell, Samuel (2007) The applicability of using ground-penetrating radar to discover and map buried archaeological sites in Hawaii. *Hawaiian Archaeology*, vol. 11, pp. 62–77.

Conyers, Lawrence B. & Goodman, Dean (1997) *Ground-penetrating Radar: An Introduction for Archaeologists*. Altamira Press, Walnut Creek, California.

Ettinger, Susanne, Manville, Vern, Kruse, Sarah & Paris, Raphaël (2014) GPR-derived architecture of a lahar-generated fan at Cotopaxi volcano, Ecuador. *Geomorphology* vol. 213, pp. 225–39.

Gaffney, Chris, Gaffney, Vince, Neubauer, Wolfgang et al. (2012) The Stonehenge hidden landscapes project. *Archaeological Prospection*, vol. 19, pp. 147–55.

Goldberg, Paul & Mcphail, Richard I. (2006) *Practical and Theoretical Geoarchaeology*. Blackwell Publishing, Malden, Massachusetts.

Gomez, Christopher, Kataoka, Kyoko S. & Tanaka, Kenji Tanaka. (2012) Large-scale internal structure of the Sanbongi Fan–Towada Volcano, Japan: Putting the theory to the test, using GPR on volcaniclastic deposits. *Journal of Volcanology and Geothermal Research*, vol. 229, pp. 44–49.

Goodman, Dean, Nishimura, Yasushi, Hongo, Hiromichi Hongo & Higashi, Noriaki (2006) Correcting for topography and the tilt of ground-penetrating radar antennae. *Archaeological Prospection*, vol. 13, no. 2, pp. 157–161.

Jewell, Chris J. & Bristow, Charles (2006) GPR studies in the Piano di Pezza area of the Ovindoli-Pezza fault, central Apennines, Italy: Extending palaeoseismic trench investigations with high-resolution GPR profiling. *Near Surface Geophysics*, vol. 4, no. 3, pp. 147–153.

Rapp, George & Hill, Christopher L. (2006) *Geoarchaeology: The Earth-science Approach to Archaeological Interpretation,* 2nd Edition. Yale University Press, New Haven, Connecticut.

Trinks, Immo, Johansson, Bernth, Gustafsson, Emilsson et al. (2010) Efficient large-scale archaeological prospection using a true three-dimensional ground-penetrating radar array system. *Archaeological Prospection*, vol. 17, pp. 175–86.

Waters, Michael R. (1992) *Principles of Geoarchaeology, A North American Perspective*. The University of Arizona Press, Tucson, Arizona.

Glossary of Common GPR Terms

air waves: recorded radar waves visible in reflection profiles, which move in air from the transmitting antenna to a reflection surface and then back to the receiving antenna. These waves are often straight, but can be of various shapes depending on how the antennas are moved during recording in relation to the reflective surface. The reflection surfaces can be walls, trees, stones, vehicles, above ground utility lines, and any other objects that can reflect radar waves. They are more common with lower frequency antennas where radar energy spreads out from the transmitting antenna.

amplitude: a measure of the "strength" of radar waves recorded by GPR systems. These values are recorded as dynamic range of digital values that define each sine wave recorded. Variations in wave amplitudes are a function of differences in velocity of traveling waves as they cross bounding surfaces that reflect energy, with the greater the velocity contrast, the higher the reflected amplitude.

amplitude maps: common maps produced by resampling the digital values of amplitudes recorded from interfaces in the ground. They are often referred to as "time-slice" maps or "depth-slice" maps, as they are produced from slices of ground defined by wave-recording times or depth. Most often they are generated over a "thickness" of material in the ground, such as 5–10 ns or 20–40 cm. They can also be constructed from only one distinct plane. These maps can also be produced to follow specific horizons that vary in their depth in the ground.

antennae: in GPR these electronic devices transmit radio waves. They can be of various shapes and sizes to generate different frequency waves, with larger antennae usually producing lower frequency (longer wavelength) waves. Electrical pulses are applied to an electrically conductive material, which depending on their shape, size, and other electronic components, generate electromagnetic waves that propagate outward. They are often used in pairs, with one antenna transmitting with the other receiving and recording waves produced from reflections off interfaces in the ground or other surfaces.

attenuation: the weakening and general reduction in the strength of radar waves as they move through a medium. In the ground this occurs when waves propagate through electrically conductive or magnetically permeable materials. Weakening also occurs as propagating waves, moving in a conical transmission pattern, spread over a greater amount of ground, creating additional reduction in strength.

background removal: a processing method that can remove an average waveform from all reflection traces recorded in the ground. The method generates an average wave by

Ground-penetrating Radar for Geoarchaeology, First Edition. Lawrence B. Conyers.
© 2016 John Wiley & Sons, Ltd. Published 2016 by John Wiley & Sons, Ltd.

sampling all waves within a given number of sequential traces in a profile, which occur at the same recorded times. That average wave can be the "background noise," which was recorded from external radio transmissions or system noise. This composite trace is then subtracted digitally from each trace in a profile, hopefully leaving only those waves that were recorded from within the ground. The number of sequential waves used to compute the background average trace can be programmed, or all traces in a profile used as a default.

band-pass filters: the removal of programmed frequencies from recorded radar waves. Low-pass filters retain all frequencies below a certain value in megahertz. High-pass filters allow values above a programmed frequency to remain, removing lower values.

bow tie: slang term for a reflection feature that looks in reflection profiles like a vertical neck-tie, knotted in the middle. They are usually products of reflections generated from steep-sided trenches, canals, or channels where the conical shaped radar transmissions produce waves that are reflected off buried features in front of and behind moving antennae.

calibration, of GPR systems: precollection settings applied to the digital controls that determine the number of traces recorded within certain time windows and with a programmed wave resolution. A variety of settings are necessary for optimum data collection that are dependent on the depth and resolution desired, the amount of ground to cover, frequency of antennae used, and the physical and chemical composition of the ground.

cation exchange capacity (CEC): a measurement of the number of exchangeable positive ions that are attracted to and retained by materials in the ground. Ground that is electrically conductive has a high CEC, which attenuates propagating radar waves.

cell phone, frequency: important as these are generators of background noise for GPR. They generally range from 800 to 1800 MHz, with some lower frequencies in Europe. The newer 3G and 4G frequencies are in the higher frequency range. When using GPR antennas in the 500–1200 MHz range, interference from these devices can be a problem. That cell phone noise must be removed from data using post-acquisition filtering programs common in most software programs. All cell phone and personal communication devices will produce extraneous electromagnetic interference when transmitting, but usually not when only in receiving mode.

center frequency: the nominal frequency of antennae common in GPR. Most GPR antennae are defined by these center frequencies, but are really transmitters and receivers from a "broad band" of frequencies, often within one "octave" of the center (one-half to two-times). For instance, a 400 MHz center frequency antenna is really transmitting waves from the one octave on either size of 400, which is about 200–800 MHz.

coefficient of reflection: a parameter that describes how much energy of a propagating radar wave is reflected at an interface. The greater the velocity change, the higher the coefficient of reflection and the higher the recorded wave amplitudes.

conductivity, electrical: an intrinsic property of a material that quantifies how it allows the flow of an electrical current. The inverse is resistivity. High electrically conductive materials in the ground have a high cation exchange capacity, and absorb radar energy, creating attenuation.

coupling, of radar energy with the ground: a relative measurement of how well transmitted radar waves move across the ground–air interface to propagate into the ground. Variations in coupling can be caused by the constituents of surface materials, the placement of the antenna on the ground, the amount of tilt of antennas, the distance of the antenna off the ground, and other factors. Good coupling means radar waves have moved into the ground and are being transmitted to depth. Coupling variations along an antenna transect create anomalous reflections in GPR reflection profiles, and can distort GPR images.

data processing, post-acquisition: digital software methods that modify GPR reflection data after they have been acquired to adjust the reflections in some ways prior to display and interpretation. These methods can be vertical and horizontal axes adjustments, filtering of frequencies, gaining of reflection amplitudes, and many other methods used to overcome noise, distortion, and other common GPR variations.

deflections, of waves positive and negative: the amplitudes of recorded sine-waves collected in traces with GPR. All recorded waves have positive and negative deflections, as when paired they produce a complete wave, the product of one reflection from one interface.

distance markers (fiducial marks): manually input digital marks in recorded reflection profiles that mark distance along transects or that mark some other location. They can be placed in the data string, or as separate files, depending on the radar system used.

electromagnetic energy: energy propagated through space or a material, which is the cojoined electrical and magnetic waves. GPR waves are electromagnetic, classified as radio waves (defined only by their frequency). Other electromagnetic energy types, not applicable to this book, are infrared radiation, visible light, ultraviolet radiation, X-rays, and gamma rays.

fiducial marks: (see distance markers).

filtering: any digital process that modifies the radar waves collected, removing some portions of the data, and enhancing others in some fashion.

frequencies, of antennae: the rate at which a wave vibrates, measured per second in values of hertz. The higher the frequency, the shorter the wavelength generated. Most GPR antennae produce an electromagnetic field that creates propagating waves that vibrate in the 10–1200 megahertz (MHz) range. One megahertz is 1,000,000 oscillations per second.

gains, applied to radar waves: a parameter, or series of parameters that change the intensity, or amplitude of waves recorded by an antenna. These are numerical values often applied to wave amplitudes during collection, or during post-acquisition processing to enhance the amplitude of waves generated from deeper in ground, which have undergone attenuation, and therefore are weaker than shallower generated waves.

hyperbola fitting: a method where a computer-generated hyperbola is "fit" to one visible in a profile in order to estimate velocity. The shape of the axes of hyperbolas is a function of the velocity that waves move to and from buried "point source" reflection surfaces.

hyperbola, as in describing a reflection: shape of reflections generated from "point sources" in the ground, caused by the spreading of transmitting radar energy as it moves deeper in the ground from a surface antenna.

isosurfaces: a computer-rendered three-dimensional surface that in GPR studies can be used to visualize on the computer a reflection surface from a buried feature in the ground.

megahertz (MHz): unit of measurement of frequency common in GPR antennae, which are units of the oscillation of waves. Equal to one million hertz. One hertz is one oscillation per second.

migration: a post-acquisition processing method that adjusts reflections recorded in the ground for velocity distortions and the spherical spreading of transmitted energy. Usually in GPR it is used to "migrate" the axes of point-source reflection hyperbolas back to their source, which is the apex of the hyperbolas. Can also, less commonly, be used to adjust the geometry of steeply dipping planar reflections.

modeling: a method used in GPR to create a visualization of what reflection profiles would be generated by buried materials and objects. Produced by inputting known or inferred electrical and magnetic parameters and shape of buried materials in the computer and then generating simulated wave paths to produce artificial two-dimensional reflection profiles.

multiple array systems: using multiple transmission and recording antennae placed together in order to record a variety of reflected waves in the ground from a three-dimensional unit of ground. Usually each antenna transmits and then records its own reflections from below. Other arrays record the reflected and refracted waves from one transmitting at multiple receiving antennas. More complex arrays transmit and record from all antennae joined together. All reflections recorded in array systems need specialized software to place reflection traces into a three-dimensional package of ground for processing and visualization.

nanoseconds (ns): the time used to record the two-way travel times of radar waves. These are units of one billionths of a second.

noise: any unwanted waves recorded during GPR collection. Most commonly they are background radio transmissions, but could be internal system-generated waves or air waves, to name a few.

point-source: a discrete object in the ground that produces a hyperbolic-shaped reflection. These are often rocks, pipes, objects, or any aerially limited reflection surface.

radar: an acronym, which has now become a word in its own right, which began to be used in 1942 for reflected radio waves used for detecting objects in the air. It stands for "radio detection and ranging." This acronym replaced the British acronym RDF that originally stood for "radio direction finding."

radio interference: any background noise within the frequency of GPR antennae.

ray paths: inferred pathways that individual radar waves move in the ground or in air.

reflections: other than the obvious definition a wave being reflected from a surface, it is also commonly used as slang in GPR and seismic wave interpretation for a visually continuous planar surface visible in a reflection profile.

relative dielectric permittivity (RDP): a complex equation that produces a quantifiable value for different materials through which radar waves move, referred to as a dielectric constant by some GPR practitioners. It calculates variations in the propagation of an electromagnetic field, depending on changes in the properties of various materials in the ground. It is used here as a proxy measurement for the velocity of radar waves moving in the ground. RDP is defined as one for radar energy moving in a vacuum, with greater values of RDP calculated for slower velocities. The highest RDP for GPR is 80, for radar waves moving in fresh water.

resistivity, electrical: an intrinsic property of a material that quantifies how well it allows the flow of an electrical current. The inverse is conductivity. High electrically resistive ground materials have a low cation exchange capacity (CEC), and readily allow the passage propagating radar waves.

shields, used in antennas: any material that impedes or destroys propagating radar waves. These can be as simple as a copper plate with attached resistors, which has a high electrical conductivity and absorbs all radar waves, which are then destroyed before they can propagate. Other materials (such as "stealth" compounds) can also be used to stop the propagating waves moving in certain directions. In GPR antennae they are usually placed to stop energy propagation upward and to the sides, so that energy will move only downward. They have less effectiveness in lower frequency antennas, and are not used at all in very low (50 MHz or lower) antennae because of their size.

spreading of radar waves: movement of waves from a surface antenna in a generally conical shape, with the apex of the cone at the surface antenna. The conical radiation pattern produced by most GPR antennas is elongated in the direction of antennas movement, if the paired transmission and recording antennas are placed perpendicular to the transect (the usual way antennae are moved for most applications).

stacking: word that can be used to describe how multiple reflections generated from layered buried interfaces are placed vertically to generate a reflection trace. Also commonly used in a somewhat different way to describe a method used to average traces during collection, or in post-acquisition processing to generate a less noisy or distorted reflection profile. Stacking during trace averaging usually takes one or more traces in front and behind one trace to average, storing the averaged trace in place of the trace in the middle. For instance, a "stack" of seven traces would average the amplitudes of three traces in front and three behind one trace to be recorded. The stacking values must be odd numbers of traces and the process is then continued for all traces in a profile, in a moving average.

survey transects: any line along the ground surface that an antenna moves. Often they are usually linear if collected within a grid, using a Cartesian coordinate system to define their location. But they can be placed in any orientation or geometry if antennae are moved around obstacles, or placed in a way to optimize how reflections are recorded from buried interfaces.

trace: a digital recording of waves recorded at one spot on the ground. Usually composed of multiple reflections recorded within a "time window," where all waves are "stacked" into one composite waveform. Traces can be analyzed individually to help define reflection at one location, but are most commonly "stacked" together sequentially along a survey transect, to generate a reflection profile.

time window: a period of time, measured in nanoseconds, in which a GPR system is programmed to record waves that intersect the receiving antenna.

travel time: usually the "two-way" time that is measured from when a radar wave leaves the transmitting antenna, moves through a medium, and is then received and recorded at a paired receiving antenna. Can sometimes be "one-way" if antennas are separated and certain types of velocity tests are being performed, or packages of ground are being studied by separating antennae in some other study method.

velocity analysis: any method that estimates the travel speed of radar waves in a medium. Important so that travel times can be converted to depth, or distance, in GPR studies.

wide-band: term to describe most GPR antennae that transmit and receive from more than one frequency. For most GPR antennae this is usually one octave around the "center frequency."

Index

accumulation of clay, 63

adobe, 57, 58, 114, 121–124, 136

adobe melt, 57, 58, 114, 122, 124, 136

aeolian deposition, 3, 6, 7, 8, 9, 21, 38, 39, 40, 50, 52, 68, 69, 74, 76–89, 122, 133, 135

aerial photos, 95, 96

aggradation, 48, 68

aggrading sediments, 40, 47

agricultural settlements, 115

air waves, 127, 139, 142

alluvial fan, 47, 57, 58, 59, 60, 61

alternating coarse and fine sediments, 97

analysis of sedimentary environments, 1–4, 9, 10, 19, 21, 37, 46, 48, 61, 64, 69, 85, 93, 100, 118, 119, 130, 132, 133

ancient environments, 1, 4, 6, 7, 9, 10, 21, 36, 48, 49, 51, 52, 61–64, 68, 70, 71, 72, 75, 76, 92, 94, 119, 132

angular boulders, 94, 95

anoxic environments, 93

antennas, 2, 5, 6, 35, 39, 71, 78, 82, 86, 87, 94, 96, 108, 116, 126, 139–143

anthropogenic units, 2, 4, 38, 40, 43, 48, 52, 54, 63, 76, 88, 100, 114, 117, 118, 133, 135, 136

aquitard, 72

architectural settlements, 75, 114, 115, 119, 125–130, 136

architectural fill, 119

antennas, array of, 130

arroyos, 65, 66

Ashkelon, Israel, 75, 91, 92

attenuation, 4, 14, 16, 17, 27, 32, 64, 68, 76, 82, 83, 93, 139, 140, 141

axes/hyperbolas, 18, 20, 81, 141

B soil horizons, 65, 69

background removal, 17, 139

band width, 15

bar types, 48

bat guano, 109, 110, 112

batch processing, 14, 137

bauxite, 82, 83, 84

bays, 74

beachs, 8, 75, 76, 77–81, 83, 89, 133

beach sand, 76, 77, 79, 80

bedding planes, 12, 14, 15, 31, 48

bedrock, 4, 6–9, 50, 51, 63, 65, 69, 76, 78–84, 102, 104–110, 112, 136

beds, stratigraphy and geometry, 8, 12, 22, 38, 43, 49, 54, 71, 84, 88, 89, 93, 95, 98, 116, 122

bentonite, 64

biological mixing, 63

bioturbation, 7, 52

boat collection, 93

bog butter, 70

bogs, 63, 70, 71, 93, 133, 135

bones, 65, 88, 89, 102, 115, 116

bow tie reflections, 40, 41, 42, 54, 140

braided rivers, 40, 41, 42, 43, 46, 47, 48, 49, 57, 61, 67

braided channels, geometry, 47, 48, 49, 57, 67

Bt unit, 68

building stone, 124

buried buildings, 59

buried living surfaces, 119

buried soils, 8, 40, 42, 43, 44, 64, 65–70, 77, 78, 101

C soil zones, 63

caliche, 64

canals, 28, 29, 40, 41, 42, 43, 54, 55, 56, 57, 65

carbonate, 63, 64, 65, 76, 78, 83, 122, 136

carbonate sand, 78, 83

caskets, 82, 83

cation exchange capacity, 64, 73, 140, 142

caves, 100, 107–112, 135

CEC, 64, 140, 142

ceiling collapse, 122

cell phones, 15, 140

cell phone, frequency, 140

cellars, 126, 130

cement, 76, 122

cemetery, 78, 82

center frequencies, 18, 23, 71, 117, 140, 143

center frequency, variations in antennas, 18, 140, 143

charcoal, 85, 87, 89, 102, 104, 105, 110, 112

chemical weathering, 3, 4, 5, 12, 18, 20, 93, 140

Ground-penetrating Radar for Geoarchaeology, First Edition. Lawrence B. Conyers.
© 2016 John Wiley & Sons, Ltd. Published 2016 by John Wiley & Sons, Ltd.